The Earth

Manfred Gottwald

The Earth

Space Travels

Manfred Gottwald
Worpswede, Niedersachsen, Germany

ISBN 978-3-662-69632-3   ISBN 978-3-662-69633-0   (eBook)
https://doi.org/10.1007/978-3-662-69633-0

For Barbara

With her, I have been traveling around the Sun for decades.

"When we see the Earth as it really is, how it is small and blue and beautiful in this eternal silence, then we see ourselves as travelers on Earth, companions on this radiant beauty in the eternal cold - companions who now know that they are truly companions."

After Archibald MacLeish (American poet and politician 1892-1982)

Prologue

Since the middle of the last century, one could feel like in a scientific paradise as an astronomically interested person. Ever larger telescopes allowed ever deeper looks into the universe; with the help of theory, we could get closer to the time of the Big Bang and new instruments, brought above the Earth's atmosphere by space travel, showed the sky in wavelength ranges that were previously inaccessible. Today's astronomical worldview is the result of an enormous gain in knowledge over the last decades, but still leaves countless questions unanswered. The universe remains a grandiose laboratory for generations of future scientists. But not only has our view to the edge of the universe sharpened, the Earth has also moved back into the center of scientific curiosity - a kind of modern geocentricity; has set in. Several causes contributed to this: space travel allowed us to explore the part of the universe closest to us, the solar system, directly on site. We recognized worlds that fundamentally differ from ours. Above all, we became aware of how unique the life-friendly environment is that our home planet offers. At the same time, people began to understand the Earth as a system, whose various components such as the Earth's interior and crust, oceans, ice and atmosphere interact in complex ways and enable the existence of life. This changed understanding of the Earth is not only taking place in our minds, it can also be directly experienced. What was hidden from previous generations can now be made visible at any time; the Earth as a colorful haven of life in an otherwise quite empty and dark universe. Anyone who deals with space travel inevitably comes across such images of the Earth. Either wonderful details become visible from close up or from great distances it reduces to a small bluish shimmering sphere, the sight of which inspires and amazes the viewer.

This book has three origins: first, a longer article by me on this topic in the German astronomy magazine *Sterne und Weltraum (Stars and Space)* as well as lectures, where I could repeatedly notice how interested listeners were captivated by a journey away from the Earth, but always looking back at it, and finally an issue of National Geographic Magazine from May 1936, in whose large-format supplement the curvature of the Earths horizon could be seen for the first time. This picture hangs in my study and constantly reminded me to write the book. *The Earth – Space Travels* describes in stages, which move further and further out into space, how our Earth increasingly appears as part of the universe. We begin our excursion in the second chapter after an introductory historical review and considerations, what we can expect on this journey at all, with the very first attempts to rise so far above the Earth's surface that one could guess the spherical shape of the Earth from the curvature of the Earths horizon. The space age was still far off at that time. This takes shape in the third chapter, when rockets reached near-Earth space for the first time without being able to already enter an Earth orbit. Now we began to see the Earth from truly great heights and also recognized the benefits for many areas of daily life, such as weather forecasting. Just a few years after the launch of the first artificial Earth satellite in 1957, weather satellites were already sending images of our planet to ground stations. This marked the beginning of a continuous development of ever more advanced satellite platforms and instruments installed on them, which now transmit enormously detailed images of the Earth's surface to us. This progress is explained in the fourth chapter. Automatic probes deliver exact and, after appropriate processing, often very aesthetic images. What they lack are personal impressions contained in such recordings. They come into play when astronauts describe their feelings to us. The photos they bring back from manned flights around the Earth are witnesses to their enthusiasm. Chapter five reports on this. While there were initially efforts to use such images scientifically, all that remains is the intention to show the viewer left behind on Earth the beauty of the Earth from above. The following sixth chapter takes us for the first time into the area where the Earth as a whole can be seen floating freely in space. It is mainly geostationary weather satellites that have been providing us with well-known views of the Earth and its weather events for more than five decades. Now the Earth also shows phenomena which can be derived from simple celestial mechanical facts such as rotation around its tilted axis or annual movement around our central star.

The sight of the Earth from another celestial body is always unique. This could be experienced for the first time when automatic Moon probes directed their cameras at the Earth from almost 400,000 km away. It had become a small disc illuminated by the changing phases. Chapter seven explains how, in the course of preparing for the manned Moon landing program, robots explored the Moon and occasionally aimed their cameras at the Earth. After the Moon had initially aroused little scientific interest after the successful completion of the Apollo program, a renaissance of unmanned Moon missions later emerged. Their views of the Earth far surpassed those of the first exploration phase.

As was the case with near-Earth orbits, the astronauts of the Apollo Moon flights contributed their personal impressions. They are discussed in chapter eight. What began with the spectacular flight of Apollo 8 in December 1968, ended already four years later in 1972. Since then, no human has set foot on the Moon again and can report on the wonderful sight of the Earth from a lunar perspective.

Recently, there has been a new stage on our journey away from the Earth. It is the Inner Lagrange point between Earth and Sun at a distance of 1.5 million kilometers, where a small, low-mass object can remain force-free and moves around the Sun in one year like the Earth. There, a satellite was placed several years ago, which mainly studies the solar wind, the constant stream of charged particles emanating from the Sun. It also has a camera that looks back towards Earth from this position and for the first time shows its constantly sunlit day side.

If one were to ask the question of the greatest success of space travel, in my opinion, the exploration of the solar system would be very high up. Interplanetary space travel made visible and tangible what astronomy from the ground could never have achieved. As a by-product, views of our home planet from entirely new perspectives were obtained. In most missions to planets, asteroids or comets, high-resolution cameras were carried that were supposed to give us first insights into these worlds. They occasionally also sent images of the Earth back to it. How and when this happened is the subject of the tenth chapter. Unlike in the previous chapters this is not done chronologically, but ordered by destinations. At the end, there are finally images from the distant regions of our solar system, in which the Earth appears only as a faintly glowing dot. From this, one can no longer recognize that billions of people share their "spaceship".

The far largest collection of views of the Earth is provided by the two stages of low-Earth orbits in chapters four and five. Both unmanned and manned missions have now brought myriad data and images or photos extracted from it to Earth. It is impossible to make an absolute selection from these. My selection in this book is intended to show the Earth system in all its facets. This includes the lithosphere, i.e., the solid surface, the world of ice in the form of the cryosphere, and the protective air envelope as the atmosphere. These three areas can also be found on other planets and Moons in the solar system - naturally characterized by other properties. So far unique to Earth is the hydrosphere, the world of water. It is also a prerequisite for the Earth's biosphere to have developed in an evolution lasting several billion years. From this, modern man finally emerged, whose transformation of the Earth's surface is now visible as the anthroposphere.

Readers may wonder why image material from US institutions is disproportionately represented. There are two reasons: Space travel has been an intensively operated field beyond the Atlantic since its inception. In addition, such ventures have always been handled very openly there, with little secrecy. In the past, their results were made known to the public in print media, today they can be found in the vastness of the Internet. NASA also strives to prevent historical documents, including photographs, from being forgotten, but to prepare them appropriately so that they can be accessed today with modern means. The former USSR certainly carried out similarly ambitious space missions in its first decades. Unfortunately, there is significantly less material from this and access to this small amount is often difficult.

The quality of the images shown here varies. Especially in earlier periods, when the cameras were often limited in their performance and there was no digital technology yet, recordings were made that no longer meet today's demands. There are sometimes projects that try to restore such old material and, if possible, even improve it. The reprocessing of Lunar Orbiter data in the seventh chapter is a nice example of this. However, when describing missions that took place a long time ago, it is also a look into the past. Recordings from this time rightly have a certain patina; they therefore appear in this book in their antiquarian appearance whenever artificial freshening would not be appropriate. The book *The Earth - Space Travels* is neither a textbook nor a picture book. Nor is it a complete photo album of terrestrial shots, obtained in more than six decades of space travel. In the archives of space agencies and scientific institutes there are still far more images of the Earth, which have been obtained with the help of satellites and probes. All the shots shown here are freely accessible and no special expert knowledge is required to access and admire them, especially with the tools of the Internet. If *The Earth - Space Travels* arouses the readers interest in retracing the journey away from the Earth outlined in it, it has certainly fulfilled its purpose.

Manfred Gottwald, Worpswede
August 2023

Acknowledgements

The Earth - Space Travels could not have been created without the kind support of various space agencies, governmental and multinational institutions, scientific institutions and commercial providers. The ventures they initiated and carried out provided the images that have taken us on a journey from the Earth far into the solar system in this book and showing us the uniqueness of our home planet. I would like to thank all of them. In particular, I would like to mention the US space agency NASA. Their rich collection of photos from space makes up the majority of this book. Also, the US agency NOAA, which among other things deals with satellite meteorology, is mainly represented with images from the early days of space travel and the geostationary orbit. Photos from the geostationary orbit are also contributed by EUMETSAT, the agency that deals with meteorology from space in Europe. Also represented in this chapter is JMA, the Japanese equivalent to EUMETSAT, as well as the Russian agency Roscosmos, whose depictions of the Earth are also attractively prepared by Vitaliy Egorov. Europes space agency ESA provides images from various stages of our journey, starting in low Earth orbit and ending in interplanetary space. From the image galleries of the Johns Hopkins University Applied Physics Laboratory, there are photos from the early days of space travel and its participation in interplanetary flights. The Japanese space agency JAXA mainly contributes the outstanding images of their Kaguya Moon mission but also some depictions of interplanetary missions. The Chinese institution CNSA, which is responsible for carrying out national space ventures, as well as the Indian Space Research Organisation ISRO, also enrich the chapter on unmanned Moon flights. In this context, I would like to thank my former DLR colleague Jian Xu, who helped me clarify the rights for the photos of the Chinese Moon probes.
The Spider Collaboration of Princeton University has provided the photo that shows the curvature of the Earths horizon from the perspective of a research balloon high above Antarctica. My thanks also go to Laura Fissel from Queens University in Canada, who mediated this illustration. Munichs European Space Imaging, a company that deals with the distribution of high-resolution satellite data, contributed the very detailed image of Munich. Finally, there is NRO, a US agency for satellite reconnaissance. Their publicly released photos show what was already possible in the early

days of space travel from the low Earth orbit with reconnaisance satellites.

Individuals are also responsible for interesting material. The photo of Explorer II with the first view of the curved Earth horizon was a particular challenge. With the support of Alexander Wittmann from United Archives, it was also possible to show this unique image document. Maximilian Reuter from the Institute for Environmental Physics at the University of Bremen has made his Meteosat images available. These have been processed in a special procedure from the original data and convey a unique natural image impression. One of these illustrations graces the title. Don P. Mitchell, a scientist who previously worked in Princeton and for AT&T Bell, and Ted Stryk from Roane State Community College have enriched the chapter on unmanned Moon missions with their Earth depictions of Russian probes from the late 60s and early 70s. Tobias Schüttler from the School_Lab of the German Aerospace Center has contributed images of a stratospheric balloon over the Bavarian Alps. They show a popular excursion area from an unusual perspective. Also conveying Bavarian flair is the photo provided by the Luitpoldgymnasium in Munich with a wide view over Bavaria. It was taken in May 2019 when the high schools P-seminar sent a balloon into the stratosphere.

In addition, there are several companies active in the space sector, scientific institutions and also individuals who were mainly involved in the success of numerous photos together with NASA and ESA. Readers will find them mentioned in the respective image captions. The free access to the data of the Copernicus Sentinel-2 mission, which allowed numerous images to be created, was also very helpful.

My colleague Eckhart Krieg deserves thanks for the effort of proofreading, which has given the texts here the final touch. Last but not least, I would like to mention my former employer, the German Aerospace Center. During my decades-long activity at DLR, I was involved in numerous Earth observation missions and was even able to make the solar system my field of research. Without this rich wealth of experience, the book The Earth – Space Travels would not have been possible.

Contents

1. Earthviews - From Fiction to Reality

"The Earth is a sphere and moves around the Sun within a year, while it rotates around its axis once in 24 hours". This knowledge is now part of basic natural science. If you are more interested, you might also know that the diameter of the Earth is approximately 12750 km and it orbits the Sun at a distance of 150 million km.

What seems trivial and logical today was not for a long time. Only since the beginning of modern scientific thinking in the Renaissance has this image of the Earth been established over the last centuries. Of the two fundamental facts - spherical shape and movement around the Sun - the spherical shape was also accepted in educated circles in the Middle Ages. Only the fact that the Earth should not be the center of the universe found no followers for religious-philosophical reasons until Nicolaus Copernicus entered the scientific stage. His heliocentric worldview could no longer be simply ignored and marked the beginning of the end of the geocentric view. Subsequently, with the invention of the telescope, it gradually became possible to learn more about our solar system. The developing physics also allowed us to understand the movement of planets and Moons within this system and finally modern astronomy placed the Sun and its planets in the correct cosmic context.

Soon after Copernicus, Kepler and Galilei had created the foundations of our worldview, descriptions of travels away from the Earth to other bodies of the solar system appeared. This includes Johannes Kepler himself with his work *Der Traum* (*The Dream*), in which the author dreams of a journey to the Moon. A little later, the French writer Cyrano de Bergerac in the middle of the 17th century embarks on his two utopian novels *The States and Empires of the Moon* and *The States and Empires of the Sun* on the way to our neighbors. Much more scientifically influenced were the reports of excursions into space by writers like Jules Verne with his works *From the Earth to the Moon* and *Journey around the Moon* or Edward Hale, who in *Brick Moon* first creates a manned artificial Earth satellite. In the first half of the 20th century, the genre of science fiction was finally established and introduced the Earth as one of countless inhabited places in the universe, the starting point and destination of interstellar space travel. The view of the distant Earth became a habit - albeit always only in utopian narratives.

What was the scientific perspective? In the introduction of the first volume of his book *Das Antlitz der Erde* (*The Face of the Earth*), published in 1883, the Austrian geologist Eduard Suess wrote "Could an observer from space, approaching our planet, push aside the reddish-brown cloud zones of our atmosphere and overlook the surface of the Earth as it rotates under his eyes over the course of a day, the wedge-shaped narrowing outlines of the continents would captivate him above all other features". This text is considered one of the first attempts to exploit an extraterrestrial standpoint to learn something about the Earth that would otherwise be difficult or impossible to discern. In a time without space travel, such considerations remained purely speculative.

In astronomy books written for the general public, depictions increasingly appeared at the end of the 19th century that visualized the view of the Earth from the Moon, for example in the book published by the French astronomer Camille Flammarion *L'Astronomie populaire*, *the Popular Astronomy*. The further one hypothetically moved away from the Earth, the more fantastic such explanations seemed at the time.

Bruno H. Bürgel undertook a very long journey in his book *Aus fernen Welten – eine Populäre Astronomie* (*From distant worlds - A popular Astronomy*), which was published in different editions from 1910 to 1958. In it there is the chapter *Die Erde als Stern* (*The Earth as a Star*), later renamed *Der Planet Erde* (*The Planet Earth*), in which the author asks himself what the Earth would look like from other planets. On a fictional journey, Bürgel travels beyond the Moon, Venus and Mars to Saturn. Each time he describes what the Earth would look like from their surfaces. Finally, in the somewhat flowery style of the early 20th century, he states "Beyond Saturn,

M. Gottwald, *The Earth*,
https://doi.org/10.1007/978-3-662-69633-0_1

your Earth is no longer known, you little man; it has sunk into the sea of light of the Sun, has slipped under its rays, and no telescope, however powerful, of the inhabitants of Uranus or Neptune could make it visible. Not yet at the borders of the solar system and already sunk and forgotten, unknown, as we know nothing of the Earths of other Suns." An interesting observation, not only with regard to the appearance of the Earth but also with respect to the existence of exosolar planets. We will encounter both statements again in the course of this book. The distances that Bürgel bridges can only be overcome with the means of space travel. When From distant worlds first appeared, Konstantin Tsiolkovsky had just published the basics of rocket technology, thus scientifically and technically paving the way to space. This path was then taken by – to name just two – pioneers like Robert Goddard in the USA or Hermann Oberth in Germany. With his book published in 1923 *Die Rakete zu den Planetenräumen* (*The Rocket to the Planetary Spaces*), Oberth made a significant contribution to the possibility of reaching foreign celestial bodies being seriously considered. It then took another 35 years until the first space probe actually left the gravitational field of the Earth and reached interplanetary space.

What does the Earth look like to us from space? Of course, it depends on how far we move away from it. The further we travel to the edge of the solar system, the smaller it will appear to us. On such a journey, we will reach distances from which we see the Earth almost as we see neighboring planets in telescopes of moderate sized from the Earth's surface. Perhaps we can even reach a distance from which, as Bürgel predicted, the Earth is barely visible. Telescopes affordable for every interested person nowadays have diameters of 15 cm and more. Either visually in calm air or with the help of fast digital cameras and suitable computer programs, surprisingly detailed views of planets can be realized. On Mars, surface structures but also large-scale storms in its atmosphere can be recognized. Jupiter shows its colored cloud bands and is orbited by its brightest Moons, which no longer appear as dimensionless points of light. Saturn's rings fascinate, which change their position relative to the Earth observer over the years. In addition, cloud bands can also be seen in Saturn's atmosphere, albeit more moderately

than on Jupiter. Even Venus no longer appears as a dazzlingly bright object; it is now even possible to detect details in its dense atmosphere with amateur telescopes.

Since Earth has approximately twice the diameter of Mars, it will appear twice as large to us from Mars as the Red Planet does from our home planet. Compared to Jupiter or Saturn, however, Earth is only about 1/10 their size, so we would see it ten times smaller from Jupiter or Saturn than the two gas giants from Earth. Just like all planets, even the largest ones, appear as very bright points of light in the night sky when observed with the naked eye, it will not be sufficient to simply look towards Earth from an interplanetary spacecraft at a great distance in order to recognize details on it. This is still possible from the Moon's orbit, where Earth appears three and a half times larger than the Moon from Earth, but beyond that, the visible Earth disc becomes smaller and smaller until it finally becomes just a faintly glowing point. We also need optical aids on the spacecraft in the form of cameras, telescopes, and recording media to produce spatially resolved images. The scales and resolutions that can be achieved depend on the properties of the optics, such as diameter and focal length, as well as the number of available pixels. A telescope with a 15 cm aperture can still separate a smallest angle of 1 arcsecond in green light. The linear dimensions this corresponds to now depend on the distance between the telescope and the object being photographed. At a distance of 400,000 km (Moon), 1 arcsecond represents a distance of 1.9 km, at 60 million km (Mars in Earth's vicinity) it is already 290 km, and at 1.4 billion km (Saturn) finally 6800 km. From the Moon's distance, therefore, a small telescope of this size could recognize details on the Earth's surface in the range of about two kilometers. From Mars, structures on Earth that measure a few hundred kilometers could be detected, and from Saturn's distance, one could just barely detect how Earth appears as a small, extended disc. Beyond Saturn, only large telescopes would show that Earth offers a richly structured surface to an observer. However, we will never transport such optics to the outer solar system.

An external observer would certainly find the dimensions of our constant companion interesting. For its dimensions, Earth has a rather large satellite; in all other cases in the solar system, the planet's diameter clearly exceeds the diameter of the respective moons. Together with the Moon, our home planet appears more like a "double planet" when viewed from the outside. We can therefore expect to be able to follow Earth and Moon as an inseparable pair far out into the solar system. However, before we venture into these areas, we must first detach ourselves from the Earth's surface in initial stages. This was not easy.

1 Earth and Moon in a image by Bruno H. Bürgel. This drawing formed the introduction to the chapter *The Earth as a Star* in his book *From Distant Worlds - a popular Astronomy* from the year 1920.

2. Close to Space - Every Beginning is Hard

Where do we want to start our journey away from the Earth's surface towards the planet Earth? It makes sense to choose a first step that does not lead too far and can actually be carried out with relatively little effort. Therefore, in the first stage, we will only try to reach heights from which one can at least recognize the spherical shape of our home planet based on the curvature of the horizon. To prove this curvature, some conditions must be met. We need an elevated viewpoint and the horizon line should be flat and not modulated by mountain or hill chains. In addition, a clear view void of clouds is necessary so that the Earth/sky boundary is clearly defined. And finally, we need a wide field of view so that the sinking of the horizon to the left and right of the stands out prominently with the highest point in the image in the middle. If one only applies geometric considerations, it naturally follows that practically at any position of the observer above the horizon, it curves to the left and right. However, when you bring a receiver like the human eye or the sensor of a camera into play, the sinking of the horizon can only be registered from a certain height. Especially with cameras, optical image errors that bend horizontal lines must be excluded. If all conditions were optimally met - which rarely happens in reality, the curvature of the Earth's horizon should present itself to the human eye from a height of just over 10 km, a perfectly photographic shot might succeed from a slightly lower distance. At a height of 10 km, the horizon is at a distance of about 400 km. The proof of the curvature of the Earth's horizon was actually a goal of scientific expeditions in the first decades of the 20th century. Airplanes that constantly moved at heights of 10 km were still rare. One could only get higher with gas-filled balloons. In 1934, the National Geographic Society, together with the predecessor of the American Air Force, set out to set a new altitude record for balloons at 24 km. The first attempt, under the name Explorer I, launched from the Stratobowl, a bowl-shaped depression in the Black Hills of South Dakota, failed. Despite the loss of the balloon and the crash of the gondola, the crew was able to save themselves safely. A second ascent, now as Explorer II, succeeded on November 11, 1935. The two-man crew included Albert W. Stevens, one of the most capable aerial photographers of the time. Explorer II reached the record height of 22 km. Among the scientific instruments carried were also two cameras, one for the view downwards, the other sideways to the horizon. Infrared material guaranteed a better view to the more than 500 km distant horizon. For the first time, images were taken that hinted at the potential of Earth images from great heights. Among them a view of the more than 200 km distant launch site in the Black Hills and even further behind the clearly curved Earth horizon. This photo was widely distributed - in 1936 it appeared as a 43 × 61 cm large supplement to the May issue of National Geographic Magazine. Also because the ascent of Explorer II was well prepared and documented, this photo is considered the first indisputable proof of the Earth's curvature.

Twenty kilometers above the Earth, one is already moving in the stratosphere. One has left behind the bulk of the mass of the atmosphere and when looking upwards, the deep blue, almost black sky is striking. This is the beginning of the "Near Space", which extends up to a height of about 100 km This value defines the upper limit of the atmosphere and the transition to space. In reality, of course, our air envelope extends even further, but at 100 km it is already so thin that an aircraft can no longer be supported by aerodynamic forces. In the years after Explorer II, interest in ever greater balloon heights faded. The emerging rocket technology promised to be able to penetrate into space and was therefore much more attractive. Today, however, stratospheric balloons are still used. They serve to carry scientific instruments to great heights, to escape the disturbing Earth's atmosphere when looking into space, or to investigate our air envelope far away from the Earth's surface in its higher floors. Such balloon campaigns are nowadays elaborate undertakings, just like in 1935 when Explorer II was launched

M. Gottwald, *The Earth*,
https://doi.org/10.1007/978-3-662-69633-0_2

from the Stratobowl for its long journey over South Dakota.

However, one can now even carry out such balloon excursions into the "Near Space" with relatively little effort. All you need is a small helium-filled weather balloon, a digital camera and a GPS module for easier finding of the payload when it has landed. The digital camera hangs well protected on the balloon and exposes at regular intervals during the ascent. The result are images, as one almost knows them from low-flying satellites: a black Sky above a deeply lying Earth surface with a curved edge. With such experiments, many schools are now inspiring their students for natural sciences. Manned balloon ascents continued after the success of Explorer II. In particular, attempts were made to move parachute jumps further and further into the stratosphere to investigate how potential pilots react to these extreme conditions. The people involved increasingly resembled astronauts. In August 1960, Joseph Kittinger exited a balloon gondola at an altitude of 31 km and returned to the Earth's surface by parachute. His record remained until 2012, when Alex Baumgartner exceeded the best mark by 8 km and just two years later Alan Eustace made a jump from almost 41.5 km. Automatic cameras captured these moments. Aircraft also reach altitudes comparable to the record of Explorer II from the 1930s. Twenty years after the launch of Explorer II, the maiden flight of the American reconnaissance aircraft Lockheed U-2 took place. It could ascend to an altitude of 27 km. The successor model Lockheed SR-71 also operated in similar regions. In contrast, the two Lockheed ER-2 high-altitude research aircraft operated by the American space agency NASA are of a civilian nature. It is now almost ninety years since the first stage into the "Near Space" was achieved. From the curved Earth horizon to an image of the Earth from the outer solar system, however, it is still a long way. First, the Earth's gravity must be overcome.

2 In this balloon gondola began the journey in November 1935 that we will undertake in *The Earth – Space Travels* far into the solar system. The two-man crew of Explorer II could peer at the Earth's horizon through small windows, while the camera to the right of the round entry and exit hatch in the center of the picture looked outside and took the photo from the Garman print edition. A second camera photographed the area directly below the gondola. Due to legal reasons the photo from the German print edition cannot be shown in the English editions. Instead Figure 5 now displays the flight track of Explorer II. (Photo: Smithsonian Air and Space Museum, CC0)

3 Balloon ascents into the stratosphere with simple means are a popular experiment at some schools. Here you can see the camera's view from 20 km altitude, as achieved by students of the Munich Luitpold-Gymnasium in May 2019 (above). The balloon was over the Inn visible at the bottom right at the time of the recording, looking north. In contrast, in 2014 a research balloon with the payload TELIS moved much higher at 36 km, which had started in Timmins in Ontario, Canada (below). Its main task was to measure the concentrations of bromine in the stratosphere. May 2, 2019 and September 7, 2014. (Photos: Luitpold-Gymnasium and DLR)

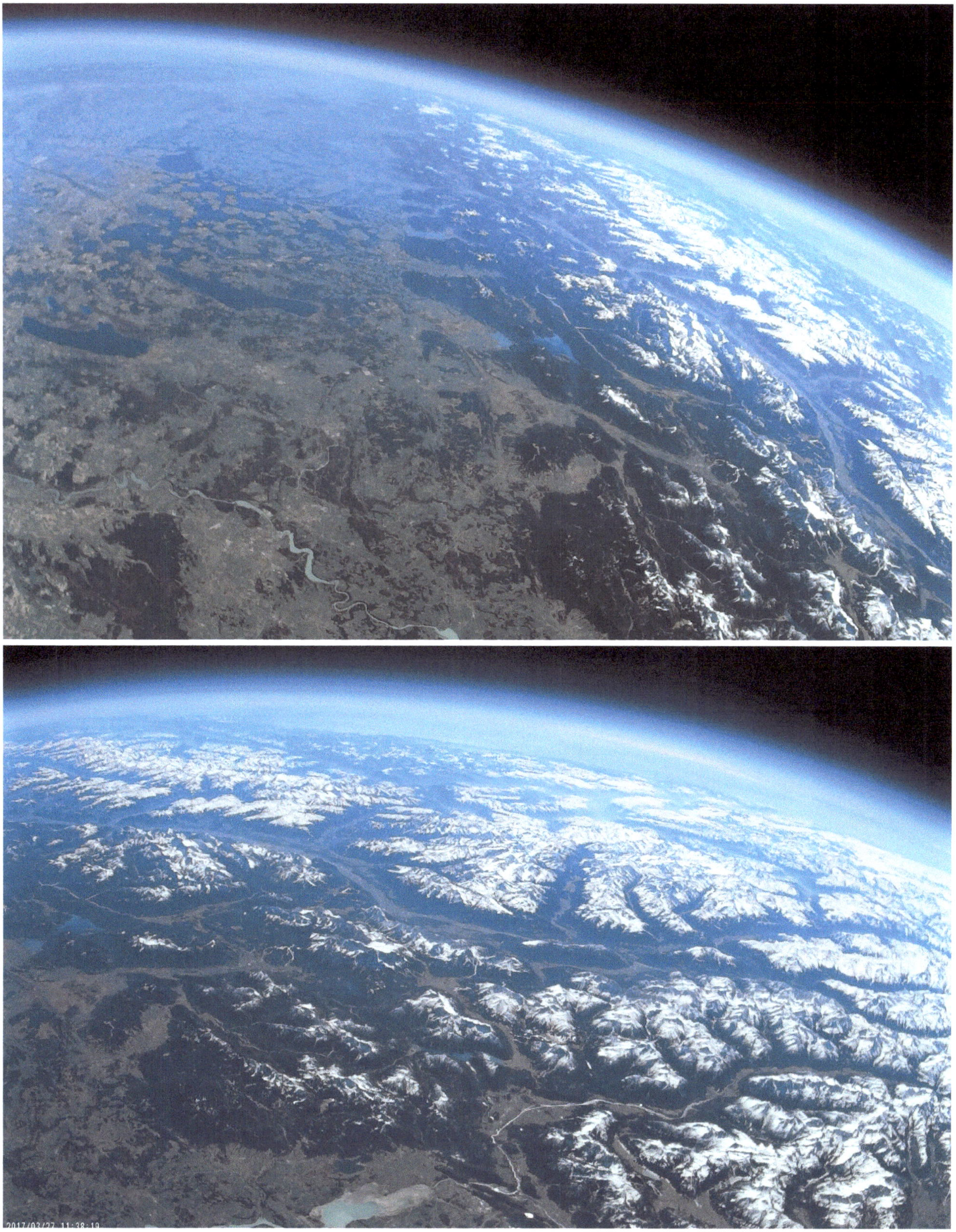

4 A balloon ascent in March 2017 from Hoher Peißenberg in Bavaria provided these photos of the Alps from a height of about 20 km. The top view looks east over the Bavarian Alps with lake Ammer, Lake Starnberg and Chiemsee. The lower picture extends far into Austria. The strong curvature of the Earth's horizon is due to the recording optics. March 27, 2017 (Photos: Team SatTec, LMU Munich)

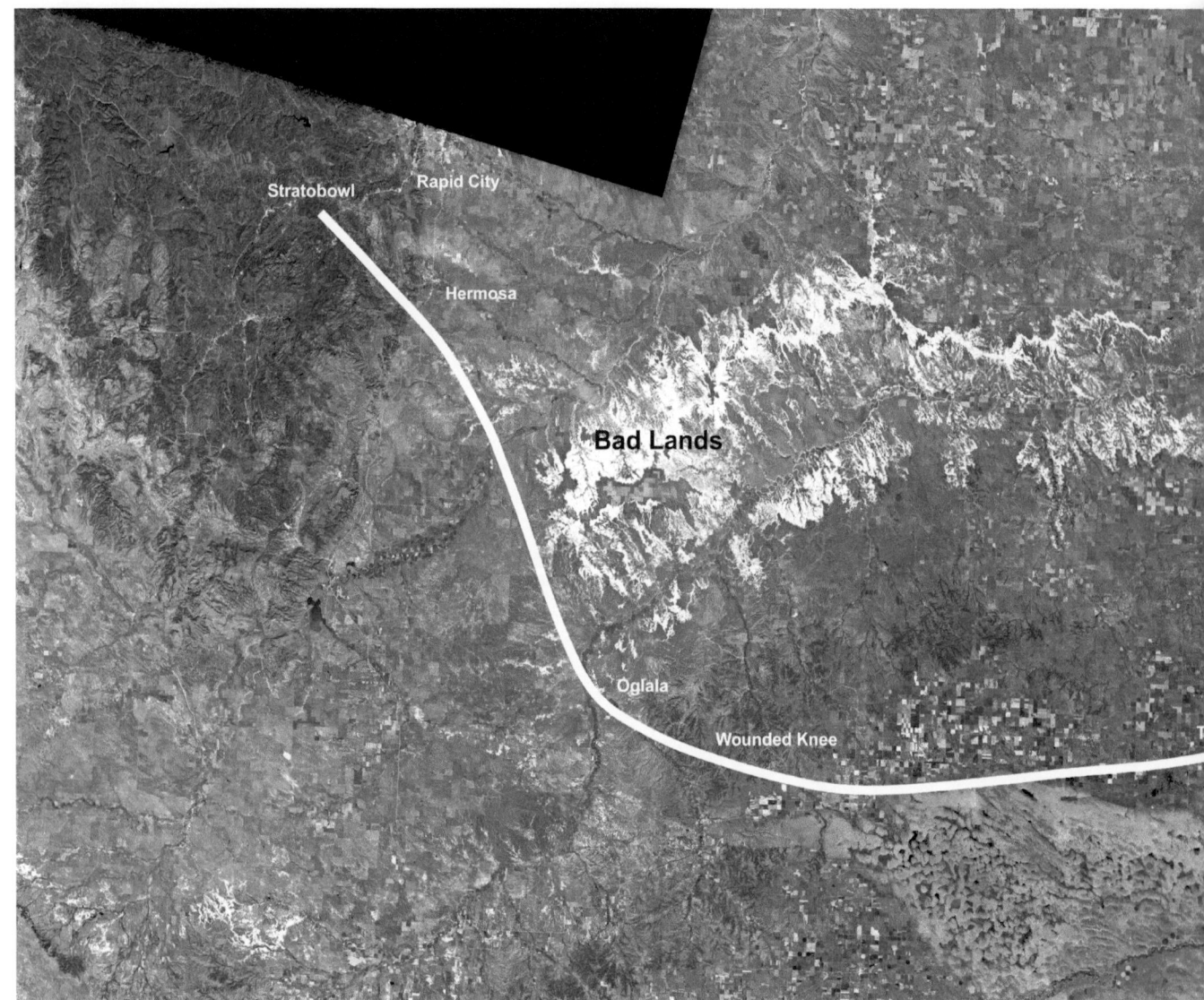

5 Due to legal reasons, the photo from the German print edition, where the curvature of the Earth's horizon could be seen for the first time in a photo from Explorer II, cannot be shown in the English print edition and in the English e-book. Instead the readers find here the flight track of Explorer II with the ascent in the Stratobowl of the Black Hills in South Dakota. The balloon drifted a long way over South Dakota eastward.

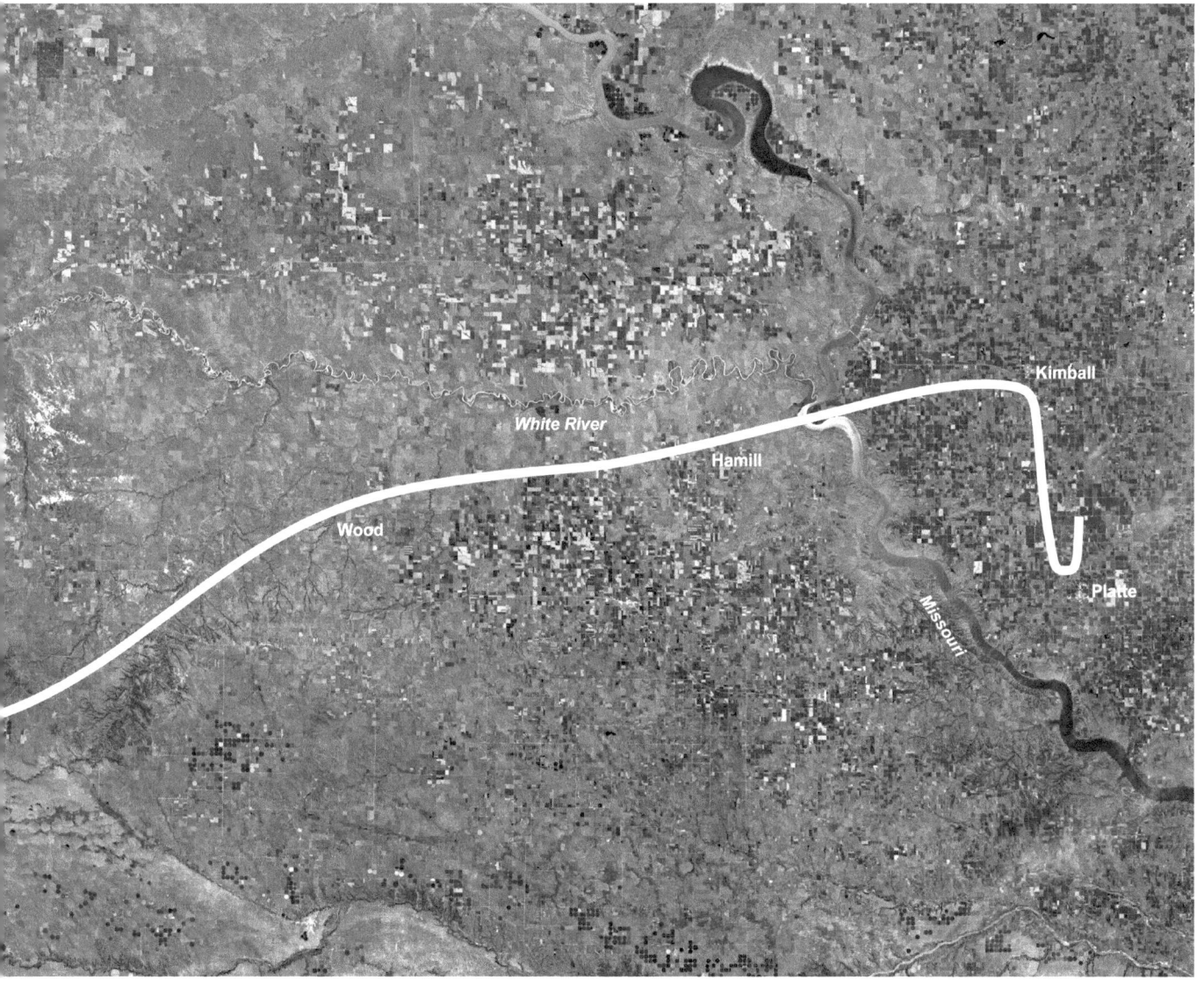

Its journey ended near Platte where it descended more than 360 km east of the Stratobowl. The whole journey of Explorer II lasted 8 hours and 13 minutes. Landsat-8, August 2023.

6 At the end of 2014, a research balloon rose high into the stratosphere over the Antarctic. The camera on its payload looked down from a height of 35 km onto the Ross Ice Shelf with the Ross Island and the volcano Mount Erebus. On the left edge of the image, one can recognize the dry valleys of Victoria Land.

The distortion of the camera optics is corrected, so here you can see the actual curvature of the Earth's horizon.
(Photo: The Spider Collaboration, Princeton University, USA)

7 At altitudes that were still record-breaking in 1935, NASA operates its research aircraft ER-2. Above is a view of the Arctic from 20 km. It was taken as part of a campaign to support the Suomi-NPP Earth observation mission, a joint project of NASA and NOAA. In the lower image, instruments for the study of aerosols, also from a height of 20 km, are being tested. The aircraft was near Flagstaff in Arizona at the time. March 2015 and November 7, 2017. (Photos: NASA, Stu Broce)

8 Joseph Kittinger's parachute jump from the stratosphere in August 1960 at the moment of jumping. We are looking from a height of 31 km at the cloud cover far below. August 16, 1960. (Photo: US Air Force)

3. Ballistic Flights - Away from Earth

The only way to approach the edge of space was to use a method that works without the presence of an atmosphere. This ruled out even higher balloon ascents or airplanes with conventional engines. The only option left was the rudimentary existing rocket technology. For the first time in the early 1940s, the first large rocket developed in Germany as a weapon, known as the V2 or A4, reached altitudes close to the edge of space at 100 km. In 1944, a climb to more than 175 km finally made a breakthrough into space on a ballistic trajectory. The development of this rocket incorporated the preliminary work of the rocket pioneers of the early 20th century. Designed with the aim of reaching as far away areas as possible to cause great damage there, this rocket reached heights that no technical device had achieved at that time. Also, the maximum speed reached by the V2, at 5500 km per hour, was significantly beyond what was usually possible at that time. The scientific monitoring in the V2 development was limited in Germany to understanding the influence of the upper atmosphere, through which the rocket moved on its parabolic trajectory, on the rocket itself; cameras were not necessary for this. It therefore took a few more years until images of the Earth from rocket flights became possible.

After the end of the war, the Allies were interested in the technology developed in Germany, then considered a large rocket projectile. The existing knowledge migrated in the form of personnel and material mainly to the USA and USSR, where it met the respective own state of knowledge in rocket development. While the progress in the USSR remained largely hidden from the public due to the Iron Curtain, the American attempts to push rocket development could be followed relatively openly. Although the military was also the driving force here, science quickly recognized the possibilities measurements at great heights, both for the Earth and for space. From the direction of space, the incidence of cosmic radiation as high-energy particle radiation was now known. Although already first detected during a balloon flight in 1912, they eluded direct measurements, as the incoming fast particles at high altitudes hit the atmosphere and only the results of these interactions can be detected on the ground. High-energy electromagnetic radiation is also absorbed in the upper layers of the atmosphere and therefore cannot be registered by observatories on the Earth's surface. At that time, speculation first arose about the existence of shorter-wave radiation in the X-ray and gamma-ray range, the existence of which should only be clearly detectable from rockets. Likewise, the existence of a high atmosphere was known, but its physics and chemistry were only inadequately understood. In addition, it was hoped that by observing weather phenomena, which should be visible in the images taken by onboard cameras, conclusions could be drawn about meteorological processes.

At the invitation of the military in January 1946 to participate in test flights with V2 rockets procured from Germany, numerous scientists registered and proposed different experiments to investigate the high atmosphere. The ride share opportunities were of course welcome opportunities for science to advance into new, hitherto inadequately known areas. This not only concerned the scientific questions, but also the fact that instruments developed for a rocket flight required technical innovations in the fields of instrument design and information transmission. The stock of V2 carriers was limited and only a few existed in complete form, but it was hoped that up to 100 functional V2 rockets could be reconstructed from the individual parts. The military had chosen White Sands in New Mexico in the southwest of the USA as the preferred test site. The first stationary engine test took place in March 1946 and shortly afterwards, on April 16, the first V2 rocket took off from its launch pad in White Sands - albeit only to a height of 6 km. By the end of the tests in 1951, a total of 67 V2 launches were carried out; half of them were failures. Even before the first V2 rockets were launched in 1946, tests were carried out in the USA with their own high-altitude research rocket, the "Aerobee". It could carry a mass of almost 70 kg to at least 90

M. Gottwald, *The Earth*,
https://doi.org/10.1007/978-3-662-69633-0_3

km. Its first launch took place in November 1947. For a long time, it was the workhorse in the high-altitude research area in different versions. Due to its limited dimensions, it could only transport small instruments, significantly smaller than was the case with the V2. It was clear that more powerful rockets of their own were needed if research at high altitudes was to be established as an independent method of investigation in Earth and space sciences, or even to reach near-Earth space. This role was assigned to a new development, the "Viking", of which about 10 copies were used until 1955 after its first launch from White Sands in May 1949. However, for a research institution, the use of the Viking was complicated and expensive; it could not compete with the smaller and cheaper rockets. The ambition to set a new altitude record drove the developers and operating teams. In September 1945, even before the start of the experiments with the V2 specimens, they managed to reach 70 km with their own small high-altitude research rocket in White Sands. Later, in December 1946, a V2 rocket equipped with instruments reached 187 km in White Sands and in August 1951 - now without instruments - 214 km. The Aerobee also moved at the beginning of its career at the peak of its trajectory more than 100 km above the Earth's surface. Later, more powerful models reached over 400 km. And finally, the Viking joined the record list for single-stage rockets in May 1954 with a maximum height of 250 km. The then altitude record was set by one of the first two-stage models in February 1949. The lower stage, a V2, carried a smaller rocket as a second stage up to 30 km, where it detached, ignited its own engine and rose up to almost 400 km.

Cameras were part of the equipment for many of the early rocket launches, often film cameras for achieving rapid sequences of images. The reason was not only scientific but also technical in nature. During parabolic flight, a rocket did not maintain its spatial orientation, but changed its position relative to the three spatial axes. In order to know at all how it was oriented, the position of the rocket was to be derived from fixed points on the ground from the rapidly taken images of the cameras. Since there was no direct transmission of the images from the rocket to the ground station, the exposed film had to be found after the end of the flight. Even a soft landing of the rocket payload was still far off;

all rockets ended their flight with an non-braked impact in the desert floor, which decelerated them from several hundred kilometers per hour within fractions of a second to zero. Hardly any camera survived such treatment, but the steel-reinforced film rolls usually remained undamaged and could be evaluated.

The first picture of the Earth from space was taken on October 24, 1946 after a V2 launch in White Sands. The rocket reached a maximum height of 105 km. After its impact, the cassette of the camera was recovered and after developing the film, the Earth was seen for the first time from above the atmosphere - a grainy piece of horizon with low-lying clouds in black and white. The quality of the recordings improved rapidly. When the images were put together into a mosaic, thousands of kilometers were now depicted. A perspective had been reached from which an entire continent could be overlooked. When 1954 the launch of a small Aerobee in White Sands was imminent, the local sky was covered. Usually, rocket launches only took place in clear weather, to ensure the visibility of landmarks in the later evaluated film and to be able to follow the rocket from the ground as long as possible. A gap in the clouds prompted the operators to carry out the launch anyway. In the images from 160 km altitude of the two 16mm film cameras, one loaded with color film, the other with black and white material, the southwest of the USA was now covered by numerous cloud fields. From the many individual images, two mosaics overlooking 4500 km were created, in which numerous meteorological phenomena could be recognized.

An interesting variant in the early rocket development from 1947 were the manned rocket planes. They served different research purposes and usually had less the goal of reaching the border to space but rather to investigate questions of aviation. However, the model with the designation X-15 can be quite compared with the activities in White Sands. The X-15 came into use only between 1959 and 1968, i.e. at a time when the space age had already begun. It was carried under the wing of a B-52 aircraft to an altitude of more than 13 km, there decoupled and ignited its rocket engine, which accelerated the X-15 until burnout after 85 seconds. After that, the rocket plane continued to climb to the peak altitude to finally land on one of the landing sites in the southwest of the USA. The greatest heights reached were even 108 km in 1963; thus, the border to space had just been crossed. On some X-15 flights, scientific measurements were taken, including photographic tests with the Earth's surface as a motif.

Although the early rocket experiments already reached into space, it remained ballistic flights. The deployment of an artificial Earth satellite was not possible. Basic physical laws still stood in the way, as even above the atmosphere the Earth's gravity acts on the rocket and accelerates it in the direction of the Earth's center. If a satellite is to orbit the Earth, this acceleration must be compensated. This is done by the centrifugal acceleration that a body moving on a circular path around the Earth experiences. It can be illustrated as follows: If the body is at a height of 200 km and circles at a speed of 7.8 km per second, or 28000 km per hour, due to gravity, he "falls" exactly as far towards the Earth in one second as his speed lifts him outward over the height of 200 km. So, after one second, the body has maintained its height - it is orbiting. The speed required to keep a satellite in orbit around the Earth is called orbital speed. It depends on the height; the value for the theoretical height of 0 km is 7.9 km per second and is also called the first cosmic speed. The further you move away from the Earth's surface, the lower the orbital speed, as the Earth's gravity decreases with distance. The speed of earlier rockets at burnout, when the rocket engine stopped operating, reached a maximum of 6000 km per hour. This value is significantly below the required orbital speed and is not sufficient to counteract the Earth's gravity.

Therefore, every ballistic rocket was denied existence as a satellite by simple physical laws. Their flight followed the same pattern: The rocket was fired almost vertically and rose steeply in the initial phase. This allowed them to quickly leave behind the lower dense layers of the atmosphere. After a certain time, the rocket was tilted depending on the range to be achieved. As soon as the burnout occurred, it continued to rise on a parabolic path to the peak height from where it fell back to the Earth's surface under the influence of gravity. When the pioneering times in rocket construction ended at the beginning of the 50s, it was difficult to guess from the images collected so far how the possibilities of imaging the Earth from space would later develop. High-altitude research rockets were no longer used for this, instead, satellites were to be used. However, they still occupied a niche and even today offer relatively simple short-term access to near-Earth space. For our journey to the edge of the solar system, the long-past mastering of the ballistic flight might have been one of the most difficult stages.

9 Right: The first picture of the Earth from space, taken from a V2 rocket over White Sands on October 24, 1946, from a height of 105 km. (Photo: White Sands Missile Range, Johns Hopkins University Applied Physics Laboratory)

10 Left: Black appearing lava fields north of White Sands from a height of 55 km from a V2. (Photo: U.S. Navy, Johns Hopkins University Applied Physics Laboratory)

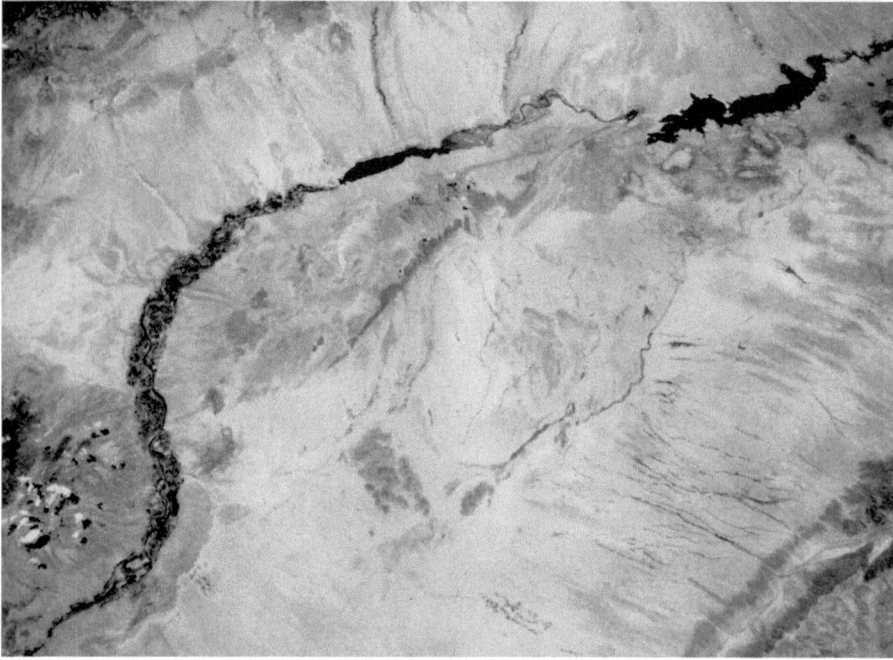

11 Left: The course of the Rio Grande with the Elephant Butte Reservoir (top right) and the semi-desert of New Mexico west of White Sands from a height of 97 km from a V2. (Photo: U.S. Navy, Johns Hopkins University Applied Physics Laboratory)

12 A view from 112 km height on New Mexico during a rocket launch from White Sands in the late 1940s. You can see the city of Alamogordo below the center of the picture and Tularosa to the left of it. Both places are connected by the Southern Pacific Railroad, whose railway line runs straight through the desert-like landscape. Towards the bottom of the picture, you can see the tracks of the Holloman Air Force Base and in the left corner of the picture, you can see the bright gypsum deposits of White Sands. (Photo: U.S. Navy, Johns Hopkins University Applied Physics Laboratory)

13 Panorama from eight individual shots of the southwest of the USA
from a height of 97 km during a V2 ascent over White Sands. The photo
spans, from south to north, a distance of 4350 km. July 26, 1948.
(Photo: U.S. Navy, Johns Hopkins University Applied Physics Laboratory)

1 Mexico
2 Gulf of California
3 Lordsburg, New Mexico
4 Peloncillo Mountains
5 Gila River

14 Mosaic of about 90 individual shots of the central and southern United States and northern Mexico. East =
Heading towards Kansas, Missouri and Iowa, South = Heading towards Mexico. The tropical storm is located
over Del Rio, Texas. The entire horizon length is 4500 km, with the apparent curvature resulting from the
assembly of individual images. This photo is considered one of those that first demonstrated the possibilities of
weather observation from space. Aerobee, October 5, 1954, altitude 160 km. (Photo: NOAA)

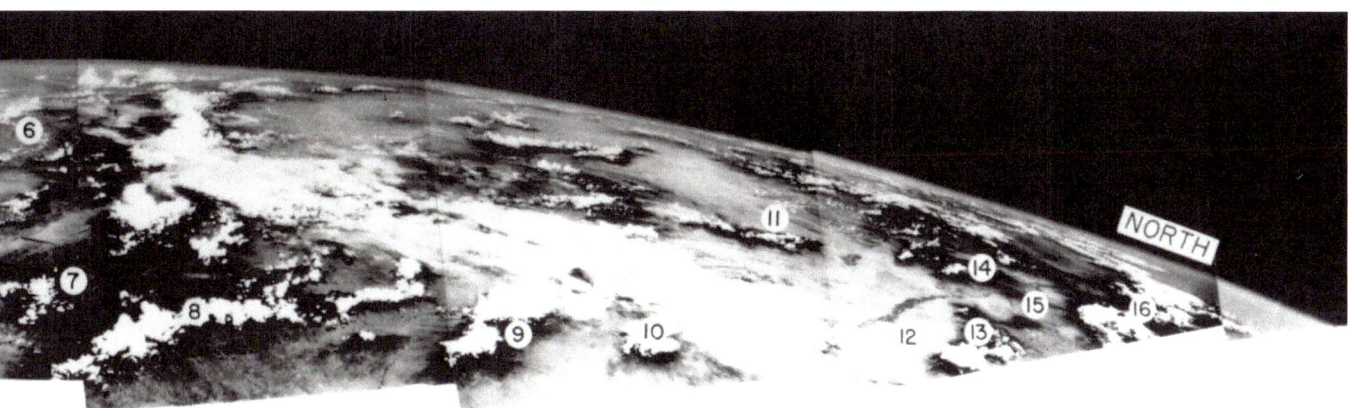

6 San Carlos Reservoir	9 San Mareo Mountains	12 Albuquerque, New Mexico
7 Mogollon Mountains	10 Magdalena Mountains	13 Sandia Mountains
8 Black Range	11 Mount Taylor	14 Valle Grande Mountains
		15 Rio Grande
		16 Sangre de Cristo Range

15 A view from an X-15 from the northwest onto Las Vegas in the foreground. To the left, you can see the Colorado River dammed up to Lake Mead and to the right, further in the background, Lake Mohave, another dammed part of the Colorado River. The rocket plane was at an altitude of about 45 km at the time of the shot. X-15, June 27, 1965. (Photo: NASA)

4. Unmanned Low Earth Orbit - Under Constant Observation

Ballistic missile flights had given us the first images of Earth from space. For our planned journey to the edge of the solar system, the distance covered was roughly equivalent to stepping out the front door when embarking on a long journey. Many of the images obtained were also more or less products of chance. The cameras on board the rockets could see the Earth's surface or the horizon, there were no more control options. Also, only the immediate vicinity of the launch site came into view for the duration of the short flight, while the entire Earth was beyond all possibilities.

In order to be able to constantly probe the globe from near-Earth space, the respective instruments had to orbit the Earth above the atmosphere. This required artificial Earth satellites. Already at times when space travel was still a utopia, its advocates speculated about permanently inhabited space stations and outposts in the universe. Satellites, as we know them today, only came into public consciousness after 1945. After initial proposals by Arthur Clarke, known primarily as a science fiction author for his work 2001: A Space Odyssey and the relocated to the USA Wernher von Braun, especially the American military conducted studies dealing with the feasibility of artificial Earth satellites. Useful applications were of course of a military nature: reconnaissance over enemy territory, improvement of weather forecasting or support of communication links. In addition, there should also be enough information for science, such as new insights about near-Earth space or research areas of astronomy inaccessible from the Earth's surface. None of the proposals were seriously pursued, not even when scientific circles increasingly took up the topic. The breakthrough came only in 1950, when a group of scientists postulated the idea for an International Geophysical Year. This idea was actually taken up by the responsible organizations and finally implemented in the years 1957/1958. The aim of the International Geophysical Year was the globally coordinated exploration of our home planet. The participating nations were to pool all their scientific capabilities in order to ultimately better understand the Earth. The plans for the International Geophysical Year also included the recommendation to develop an artificial Earth satellite and to carry measuring instruments into near-Earth orbit with its help. Five years later, in 1955, both the USA and the USSR announced that they wanted to support the International Geophysical Year with a satellite.

High-altitude research rockets are limited to a "ballistic trajectory" as they do not achieve the injection speeds required to enter a circular orbit around the Earth at a certain height. Rockets for transporting a satellite into orbit required developments from the military sector. During the Cold War, the ability to transport warheads over continental distances of many 1000 km was important. Such intercontinental missiles consist of several stages and accelerate the payload to speeds of more than 20000 km per hour, reaching peak heights of more than 1000 km on their ballistic trajectories. This is close to the requirements of a satellite launch. It is not surprising, therefore, that the launch vehicles that first carried satellites into space were developments of military rocket programs.

We already know the speed a satellite needs to orbit the Earth. At a height of 200 km, it is 7.8 km per second. If you reach higher speeds at a certain height than required for a circular orbit, it expands into an ellipse with a point close to Earth and a point far from Earth. Just as the circular orbit speed decreases with increasing height, the duration of an Earth orbit increases. If a satellite completes a full orbit of the Earth at a height of 200 km in 88 minutes, it already takes 105 minutes at a height of 1000 km. For the range of orbit heights up to 1000-2000 km, the term low Earth Orbit (LEO) has become established. Support for launching a rocket is provided by the Earth itself by taking advantage of the Earth's rotation. A rocket launched directly east at the equator receives an additional speed contribution of 460 m per second from the Earth's rotation. If it takes off at higher geographical latitudes, this component is correspondingly lower.

M. Gottwald, *The Earth*,
https://doi.org/10.1007/978-3-662-69633-0_4

This contribution results in a slightly lower required fuel quantity, which reduces the mass of the rocket at launch. The geographical latitude of the launch site has another important effect on a potential satellite launch: If you use the energetically most favorable launch direction to the east, the satellite moves without further corrections on a path that is inclined to the equator by the angle of the geographical latitude of the launch site. This takes it as far north and south of the equator as the inclination angle of the path indicates. To observe the entire Earth from pole to pole, the satellite must be given a path inclination of about 90°, which is only possible through additional maneuvers or a changed launch direction. Both variants require a higher fuel reserve and thus a higher launch weight. Once a satellite is moving in a polar orbit, the Earth rotates beneath it in 24 hours, and the satellite flies over all latitudes and longitudes. However, it can take weeks for it to appear exactly over the same place on the Earth's surface again.

But back to the 50s. The promise to launch a satellite for the International Geophysical Year was indeed fulfilled. From October 4, 1957, Sputnik-1 orbited the Earth for 98 days before it burned up in the Earth's atmosphere. The USA sent their first artificial Earth satellite, Explorer-1, into an Earth orbit shortly thereafter on February 1, 1958. In the nearly 70 years that followed Sputnik-1, more than 10000 satellites were launched into space. Of these, about 3000 are still active and in operation. Only a vanishingly small part of the satellite fleet left the Earth's gravitational field and began its activities in interplanetary space; the majority remained close to Earth. There, one of the main tasks of artificial Earth satellites has indeed become the observation and study of our home planet.

When satellites move in low Earth orbit at heights up to 1000-2000 km, the entire Earth is never visible. Although the horizon is significantly lower depending on the distance from the Earth, the Earth still covers the largest part of the field of view. The first image of the Earth, in which even a layman can recognize at least rough details, comes from a satellite in exactly such an orbit. It is TIROS-1, the first successful weather satellite. Launched on April 1, 1960, it provided an image of the western Atlantic and North America from the New England states to the mouth of the Saint Lawrence River on the same day. Of course, the image quality would

be deemed unacceptable today, but at the time, the meteorologists involved were highly satisfied. TIROS-1 carried a TV system with an imaging camera, the image produced by which could be scanned on board in strips, stored on magnetic tape, and sent to the ground station. To stabilize the satellite performed a spin motion on its orbit. Whenever the Earth moved through the field of view during the spin motion, images were taken. The alternative, recording the observed on film, would certainly have provided better quality images but at that time the stored information could not be evaluated on board. In weather forecasting, speed was required, and this could only be achieved using TV technology.

There was also a filmic approach at the same time. Intelligence agencies and the military were interested in highly accurate photos for reconnaissance purposes. The TV solution as on TIROS-1 did not meet these requirements. Only on films could the wealth of information be stored, which was registered by the telephoto lenses of the satellite cameras. But how could the films be developed? The solution was to retrieve the films in return capsules from orbit and catch them with airplanes as they glided to the ground on a parachute in the last part of their return journey through the atmosphere. The procedure actually worked; for the first time on August 20, 1960, an exposed film from the satellite launched two days earlier under the code name Discoverer-14 returned to Earth in this way. Since the mid-90s, the first spy satellites of the Discoverer series from the CORONA program are no longer classified, some of their quite remarkable results for the time are available on the Internet. TIROS-1 and Discoverer-14 were the beginning - since then a fleet of Earth observation satellites in low Earth orbits has examined our home planet from all possible aspects. 10 TIROS missions were in operation until 1966, followed by improved versions under the name ESSA. A third generation, which technically had nothing in common with the original TIROS design, but historically dates back to the early 60s, were the NOAA probes used from 1970 and still in use today, named after their operator, the American weather and oceanography agency. A second important series from NASA were the 7 Nimbus satellites, which replaced each other in orbit for a long time from 1964. They also formed the basis for the Landsat missions, which made the observation of the Earth from space known to the public through a multitude of exciting images. Similarly, on the Russian side, there were various programs, both civilian and military, with the aim of a constant view of the Earth soon after the beginning of space travel. From the 70s onwards, other countries that had by then acquired the capabilities for space travel sent their satellites for remote sensing of the Earth into space. Today, the Earth is constantly targeted by scientific, commercial, and military platforms. While the first satellites were small, manageable devices, TIROS-1 had a diameter of 1 meter and carried two simple instruments, the current missions are highly complex undertakings. ENVISAT, Europe's Earth observation satellite with ten instruments active, from 2002-2012, was one of the largest controlled unmanned structures in near-Earth space. With its solar panel extended, it measured 26 m in length. Nowadays, in the civilian sector, sensors achieve resolutions that were previously reserved for the military. Today, commercially operated satellites from a height of 800 km can detect objects of just a few decimeters in size! In the civilian sector, the early experimental ventures were exclusively dedicated to improving weather forecasting, derived from changes in large and small-scale cloud structures, gradually the desire arose to learn more about our home planet as a system. Today we know that the "system Earth" is a complex interplay of different areas is. These include the entire solid Earth with the Earth's crust containing lithosphere as the outermost layer. All water is found in the hydrosphere, which, when it freezes and takes on its solid state of aggregation, is referred to as the cryosphere. The air envelope of the Earth finally forms the atmosphere. A special feature of our planet is the biosphere, the totality of all organisms on the Earth. The interaction of these spheres often makes it difficult to make reliable predictions about important processes on Earth, such as weather and climate. Since Earth observation satellites continuously monitor the Earth, we hope that these contribute to a better understanding of the underlying interactions.

The early satellite images for weather forecasting usually registered visible light, coming from the Sun, reflected and scattered by the Earth into space, where it hit the camera of the satellite. Now visible light consists of many wavelengths from blue to red. As soon as it reaches the different "spheres" of the Earth, it is by atoms, molecules and small particles in a characteristic way, depending on its wavelength, modified. If you want to fully exploit the potential of remote sensing from space, you have to measure the visible light in separate color ranges, for example by filtering or even splitting of the wavelengths into a spectrum. In addition, the radiation emitted by the Sun contains components for which the human eye is not sensitive, but which are also changed by the Earth. This includes the energy-rich ultraviolet range below the blue light or the less energetic infrared radiation above the red wavelengths. Added to this is the radiation that the Earth also emits in the long-wave infrared due to its temperature. This is typical for the radiating body and can help to understand its properties in our case solid Earth surface, water or air better. In contrast to visible radiation, the Earth also emits its thermal radiation at night. When a satellite orbits the Earth, it spends half of its orbital period on the night side of the Earth in the dark. Devices that only register visible light, provide no results there while thermal radiation measuring sensors can continuously measure along the entire Earth orbit. Instruments which measure the radiation sent by the Earth into space, either the reflected and scattered Sunlight or the terrestrial thermal emission in different wavelength ranges, are called multispectral.

When NASA launched Landsat-1 in July 1972, then still under the designation "Earth Resources Technology Satellite 1", a new era began. Landsat-1 was equipped with a novel multispectral sensor, whose performance far exceeded everything previously existing. When it transmitted the first images of the Earth's surface, the scientists involved initially considered visible details to be image errors, before they realized that they were real fine structures. The outstanding quality of the multispectral images contributed significantly to the success of the Landsat missions which continue to this day.

Remote sensing with satellites is probably one of the most versatile scientific applications of space travel. Today our home planet is constantly probed by a large fleet of platforms with different tasks. It is no longer limited to just measuring radiation that spreads from the Earth towards space, passively, but "illuminates" the Earth from satellites actively with microwaves, which also include the radar range, to extract information from the backscattered signal. Along with the development of the increasingly sophisticated satellites were advances in the construction of image sensors and detectors and in the computer-aided processing of the resulting data. Modern instruments deliver large amounts of data that must be transferred from orbit to ground stations, where they are further processed to deliver scientific results. In the early days of Earth observation satellites, results were exclusively disseminated via print media. Today we have fast Internet connections that allow us to search worldwide for specific data or results and, if available, to retrieve them for further processing. The establishment of Earth observation from space benefited enormously from the transition from analog to digital technology. Modern image processing even allows for virtual views. In orbit, day and night alternate with half the orbital period. If an Earth observation mission lasts longer, the scenes taken over the illuminated part of the globe – just like the night shots taken with highly sensitive cameras - eventually cover the entire Earth. This can result in "day only" or "night only" images of our planet that do not exist in reality. Equally unreal are maps composed of cloud-free scenes. Usually, during an Earth orbit, a large part of the continents and oceans remain hidden from view from space due to dense cloud cover. There are areas where cloud-free shots are hardly possible at certain times. Only patient "collecting" over long time periods provides the desired information. An Earth observation satellite usually captures scenes with up to a few 100 kilometers side length. They are created by the sensors looking straight down onto the Earth's surface and scanning the area near the orbital path as the satellite moves forward in its orbit. All such images can be displayed in the computer in any projections. If you choose a sphere, you see the Earth as if you were moving far above it, even though the camera in reality only orbited only a few 100 kilometers above the Earth's surface. Ultimately, this could even enable a virtual journey to the edge of the solar system! In the preceding two chapters, it was relatively easy to make a selection of typical photos. The low Earth orbit makes it much harder for us; the archives now store an ever increasing wealth of suitable recordings. In particular, the American and European space agencies, NASA and ESA, offer interested parties free access to attractively prepared results of their Earth missions as part of their public relations work. Certainly, such a selection should show the "spheres" that uniquely characterize the Earth in the solar system, i.e. lithosphere, hydrosphere, cryosphere, atmosphere, biosphere and what human civilization has created so far, the anthroposphere. In addition, it makes sense to look for motifs that illustrate the interaction between Earth and its cosmic environment.

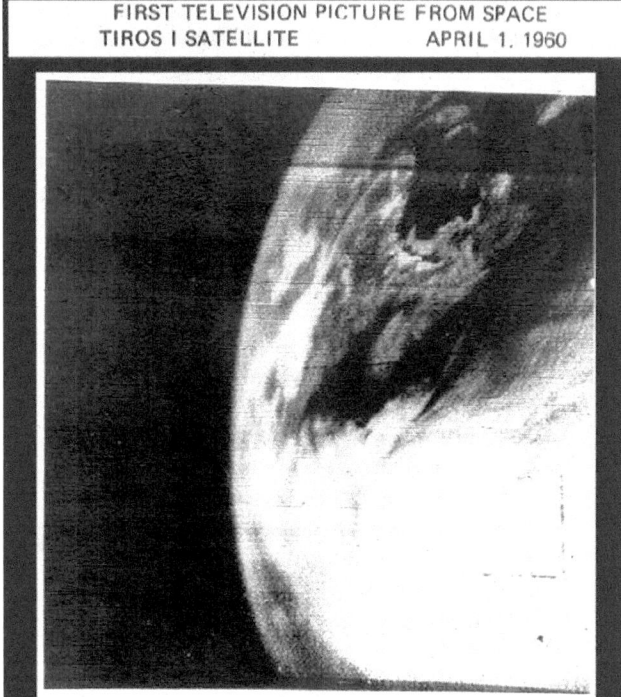

FIRST TELEVISION PICTURE FROM SPACE
TIROS I SATELLITE   APRIL 1. 1960

16 The photo that is considered the first taken of the Earth from space by a satellite. It comes from the weather satellite TIROS-1 and shows the eastern part of North America with Nova Scotia above the white cloud band. TIROS-1, April 1, 1960. (Photo: NASA)

17 Another early photo from TIROS-1 from the same day. Here you can see the mouth of the St. Lawrence River with the island of Anticosti located east of New Brunswick in the river. TIROS-1, April 1, 1960. (Photo: NASA)

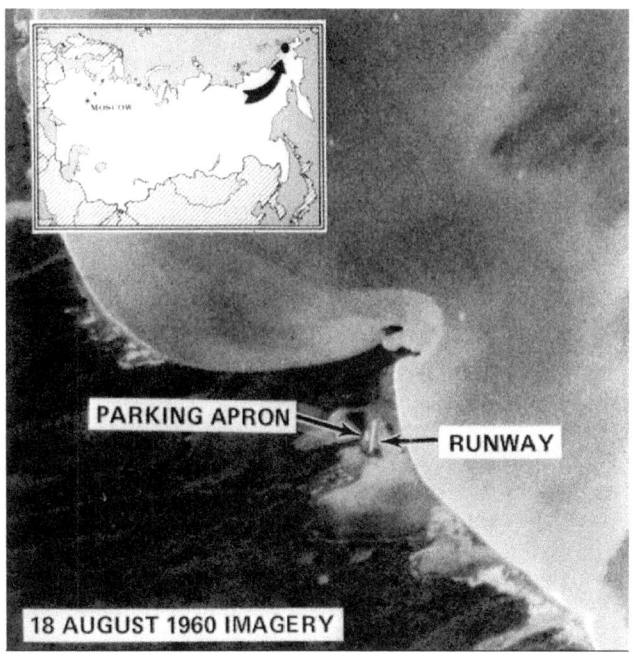

PARKING APRON   RUNWAY

18 AUGUST 1960 IMAGERY

18 The first photo of a CORONA reconnaissance satellite, developed from the film retrieved by an aircraft. Siberia, Mys Shmidta Airport, CORONA, August 18, 1960. (Photo: NRO)

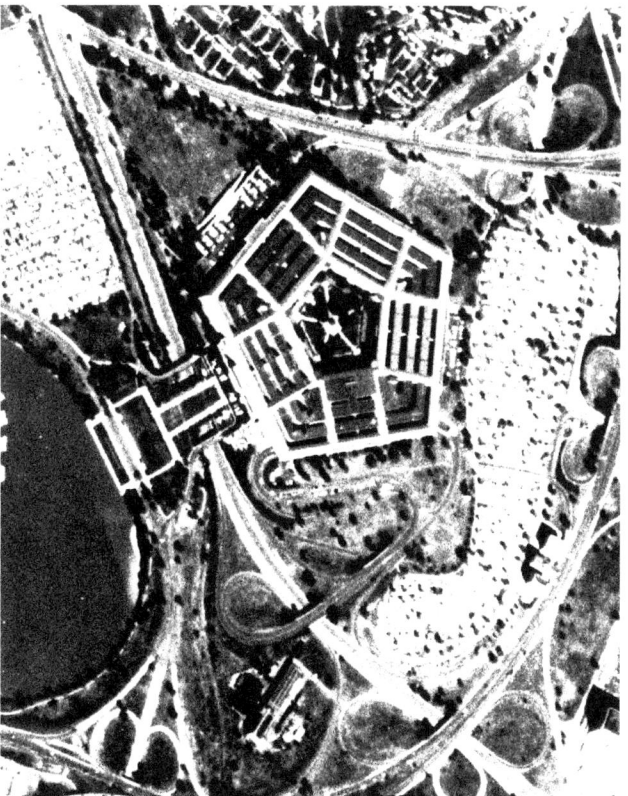

19 An example of the improvement in the performance of the cameras used on the CORONA satellites. The Pentagon in Washington appears much clearer on this image. CORONA, September 25, 1967. (Photo: NRO)

20 The image of Hurricane Anna is considered one of the first photos of a hurricane from space. TIROS-3, July 1961. (Photo: NOAA)

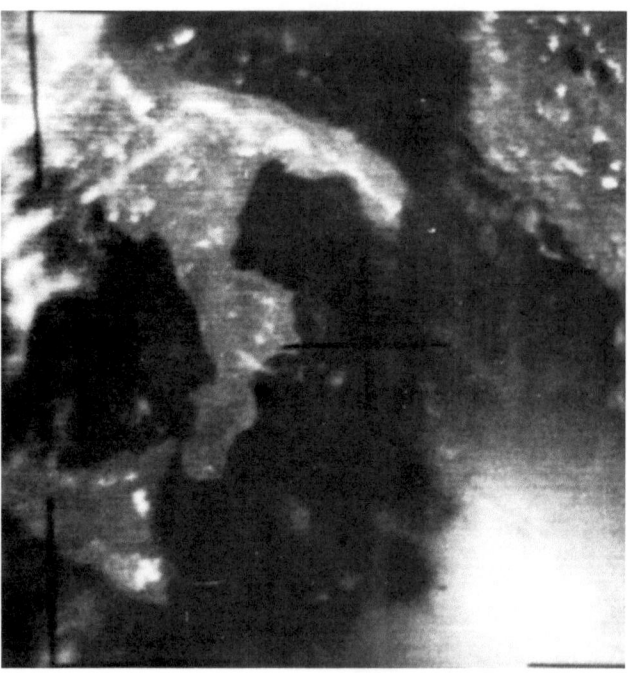

21 Southern Italy with Sicily. TIROS-7, April 23, 1964. (Photo: NOAA)

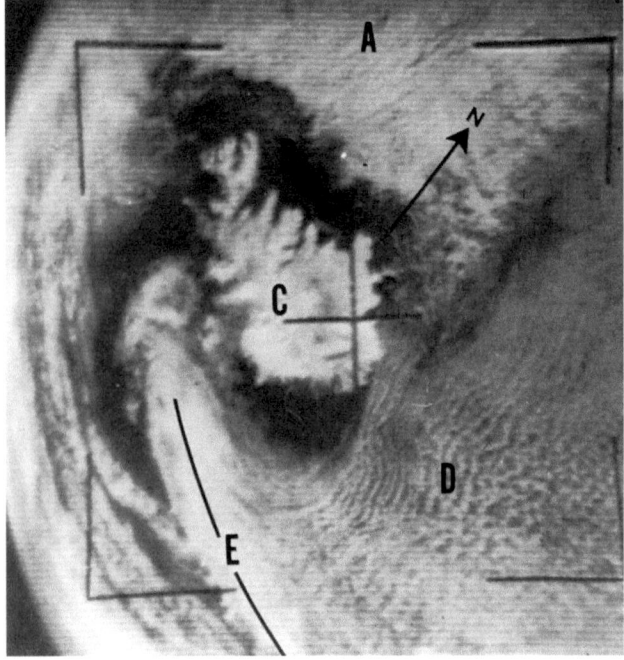

22 Iceland in a cloud gap. TIROS-9, March 22, 1965. (Photo: NOAA)

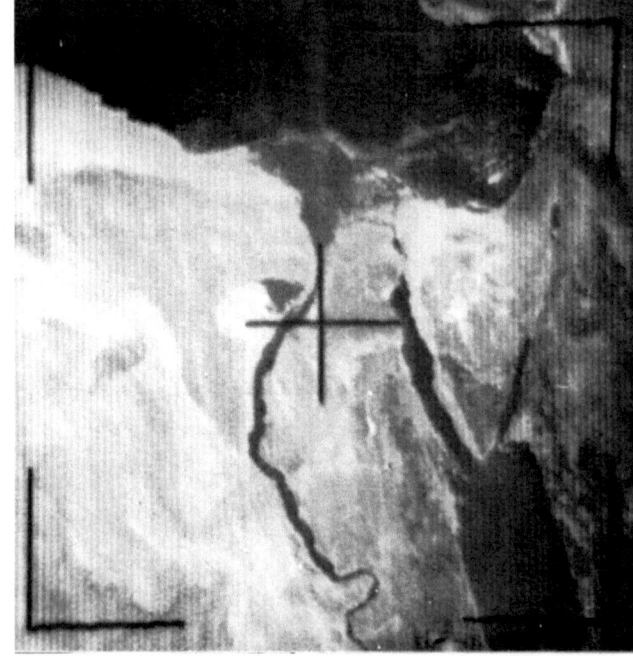

23 View of cloudless Israel and Egypt with the Sinai Peninsula. TIROS-10, September 7, 1965. (Photo: NOAA)

24 Section of the first image generated from multispectral data by Landsat-1 in July 1972 with Dallas, Texas to the left of the center of the image. It appears in the then preferred red false color image and means that data from the infrared channel were extracted. Red stands for areas with vegetation, gray and white for urban areas and rocky ground. The resolution reaches 60 m per pixel. Compared to the earlier images, this is a quantum leap. Landsat-1, July 25, 1972. (Photo: NASA, USGS)

25 A cloudless view of the southwest of the USA and northern Mexico. In the north, the snow-covered peaks of the Rocky Mountains. To the west of it, the Colorado River has carved into the reddish rock with the Grand Canyon as the highlight. White Sands is marked by the area with very bright ground to the right of the center of the image. MODIS/Aqua, November 18, 2003. (Photo: NASA, GSFC, Jeff Schmaltz, MODIS Rapid Response Team)

26 When two tectonic plates meet on their way across the Earth's surface, it depends on the type of their crust, how both behave. As the Indian subcontinent moved north over the past tens of millions of years, it encountered the Eurasian Plate. Both consist of continental crust; neither could therefore yield as heavy material and submerge back into the mantle. In the process, India pushed far into the Eurasian Plate. It led to folds that created the Himalayas as today's highest mountain range and raised the highlands of Tibet behind it. In the image, you can see the lower Indian subcontinent in the bottom left. The majority of the image is occupied by Tibet with the Himalayas as the southern boundary. MODIS/Terra, November 20, 2003. (Photo: NASA, GSFC, Jeff Schmaltz, MODIS Rapid Response Team)

27 The Alps are the result of the collision of the north-moving African Plate with Eurasian Plate and the small Adriatic Plate located between, that subducted under the Eurasian Plate. The formation of the Alps as an extensive folded mountain range dates back about 30 to 50 million years with a currently still ongoing northward movement of Africa, so the height growth of the Alps is not yet completed. MODIS/Terra, January 17, 2011. (Photo: NASA, GSFC, Jacques Descloitres, MODIS Rapid Response Team)

28 The lithosphere consists of several large and small plates, which, driven by movements in the Earth's mantle, move against each other at speeds of a few centimeters per year. They can either consist of lighter continental crust or heavier oceanic crust. Plate boundaries often run at the bottom of the ocean, where magma emerging from the Earth's mantle forms new crust material and lets the plates drift apart. One of the few places on Earth where such a mid-ocean ridge runs on the surface is Iceland. The eastern part of the Iceland belongs to the Eurasian plate and moves eastward, whereas the western part belongs to the North American Plate and moves westward. The fault zone appears in the photo as a dark, diagonal region running across Iceland. MODIS/Terra, September 9, 2002. (Photo: NASA, GSFC, Jacques Descloitres, MODIS Rapid Response Team)

29 At some points in the Earth's mantle, there exists a reservoir of hot material that can reach the surface. When a lithospheric plate moves over it, magma penetrates the plate from below, reaches the surface, and piles up there as an extensive volcano. The movement of the plate thus creates a chain of island structures made of magma rock. Such a process has been happening for millions of years in the Pacific, where the Pacific Plate moves over a "hot spot " and allows the Hawaiian island chain to form. The smaller islands located to the northwest are already inactive and are submerging back into the Pacific. MODIS/Terra, May 27, 2003. (Photo: NASA, GSFC, Jacques Descloitres, MODIS Rapid Response Team)

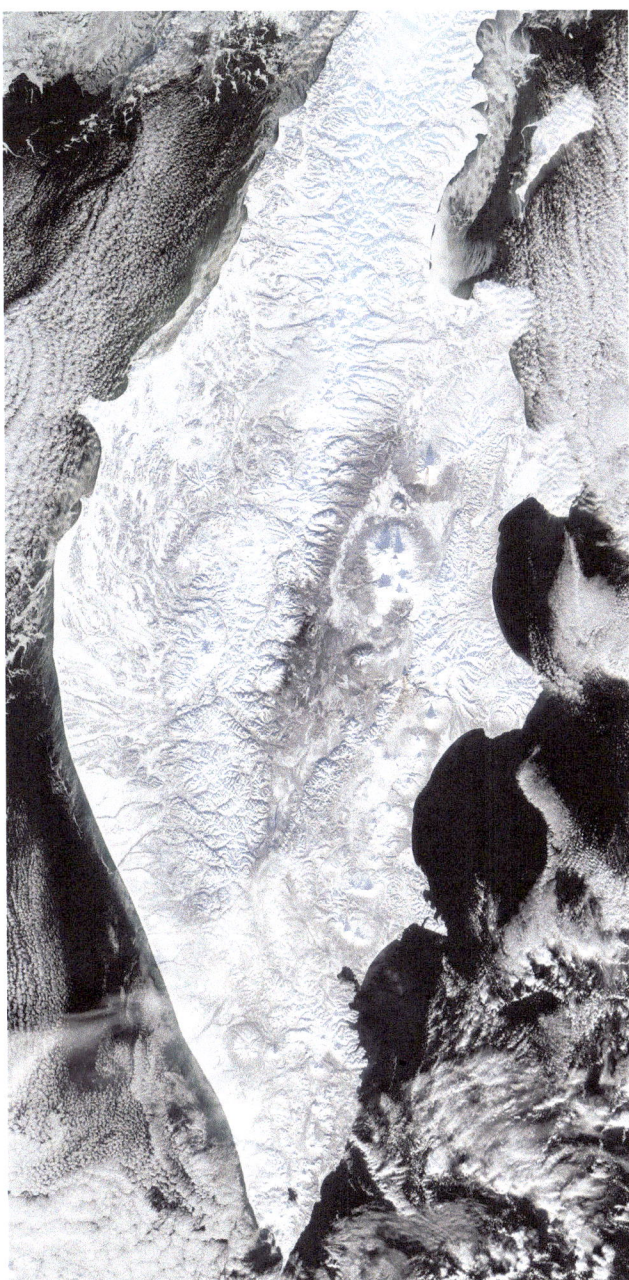

30 In the Bering Sea, the North American and Pacific Plates meet, but not directly, instead separated by two smaller plates. One of them, the Okhotsk Plate, carries a large part of the Kamchatka Peninsula, which is considered a region with the highest density of active volcanoes. The subduction of the Pacific Plate with oceanic crust leads to strong tectonic activity including volcanism. In the MODIS image, numerous volcanic cones cast their shadows on the snow-covered surroundings. MODIS/ Aqua, December 19, 2011. (Photo: NASA, GSFC, Jeff Schmaltz, MODIS Rapid Response Team)

31 Depending on the type of eruption, a volcano can eject vast amounts of the finest ash particles and transport them to great heights, where they spread widely. For global air traffic, volcanic ash poses a great danger as it can penetrate the engines and electronic systems of aircraft during flight and seriously damage them. The eruption of the Eyjafjallajökull volcano in Iceland began in March 2010 and lasted more than two months. During this time, enormous amounts of ash were emitted and transported towards Central Europe, paralyzing air traffic there. MODIS/Terra, May 11, 2010. (Photo: NASA, GSFC, Jeff Schmaltz, MODIS Rapid Response Team)

32 One of the most impressive natural monuments on Earth is the Grand Canyon in Arizona. Over more than 400 km, the Colorado River has carved into the rock of the Colorado Plateau; in some places even almost 2000 m deep. The lowest rock layers are up to 2 billion years old while the youngest, aged 270 million years, are found at the top edge of the canyon. Many of these rocks result from deposition in shallow seas or at their edges over the course of Earth's history. The formation of the Grand Canyon began 70 million years ago when the Colorado Plateau gradually rose up to 3000 m during the mountain formation in the western North American continent, which led to the creation of the Rocky Mountains. The current appearance of the canyon is estimated to be 5-6 million years old. The Sentinel-2 image of the Grand Canyon spans about 230 km in east-west direction and shows its entire course from Lake Powell in the east to where the Colorado River meets Nevada in the west. Copernicus-Sentinel-2, December 3, 2021.

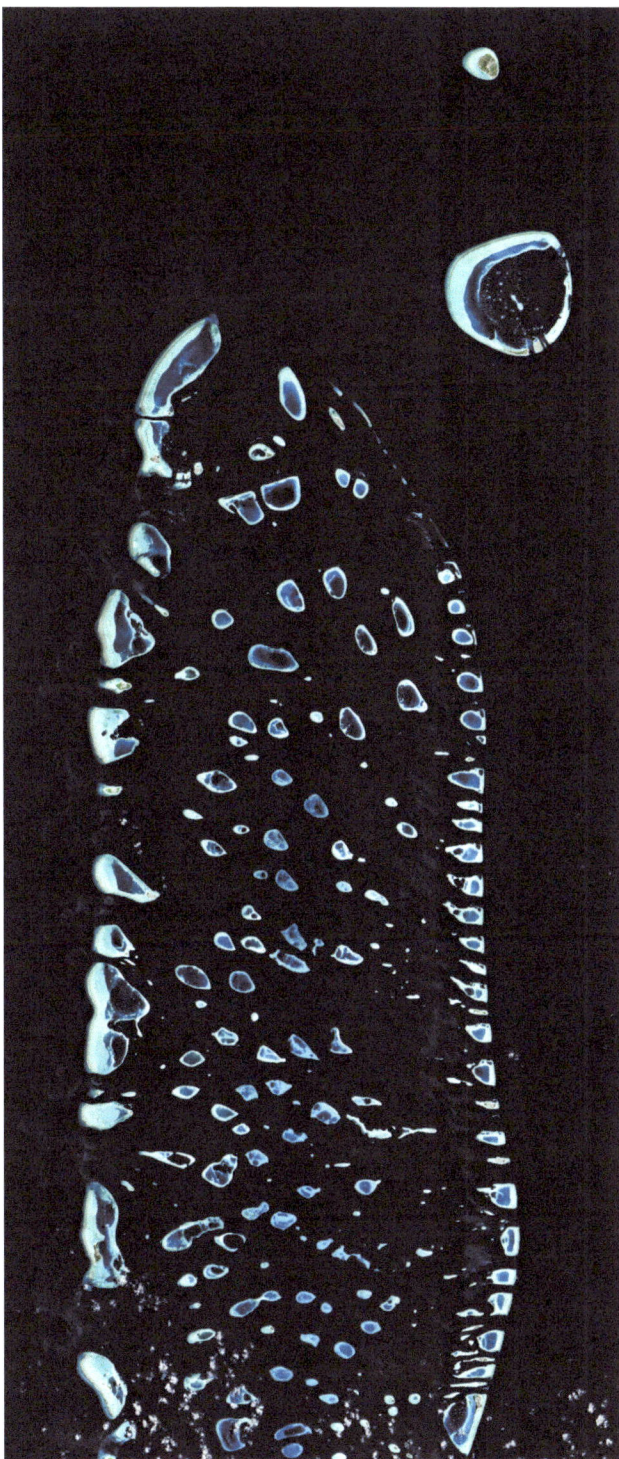

33 Earth is the only planet where liquid water exists on the surface. From Low Earth orbit, visual observers find regions particularly interesting where water meets land - either on the coasts or inland at lakes and rivers. This is especially true for the Bahamas off the southeast coast of the United States with its turquoise shallow waters and the abrupt transitions into the deep ocean. East of the Andros island group, one finds deep water, known as the "Tongue of the Ocean". At its southern end, in the shallow waters of the Great Bahama Bank, sand dunes from the residues of eroded coral reefs form interesting patterns. Landsat-8, April 4, 2020. (Photo: NASA, Joshua Stevens, USGS)

34 About 700 km south of Sri Lanka in the Indian Ocean are the Maldives, a nation consisting of about 1200 coral islands. These islands are oval atolls, formed from a sunken volcano and surrounded by the upward-grown coral reef. The Ari Atoll shown here is located in the west of the island group with an extent of 90 km × 30 km. Copernicus Sentinel-2, April 12, 2019. (Photo: contains modified Copernicus Sentinel data (2019), processed by ESA)

35 The Amazon forms the largest river system on Earth. After its origin in the Peruvian Andes, it reaches its mouth after 6500 km and releases a fifth of the Earth's freshwater into the Atlantic Ocean. In its estuary, the Amazon has a width of 200 km. Its sediment-laden water is still far into the Atlantic Ocean, recognizable by its color. MODIS/Terra, June 10, 2002. (Photo: NASA, MODIS Rapid Response Team)

36 The dumping of large amounts of sediment in a shallow river delta is also shown in the picture of the mouth of the Ganges coming from the west and the Brahmaputra arriving from the north into the Bay of Bengal. The delta is characterized by extensive mangrove swamps, the Sundarbans. MERIS/ENVISAT, November 8, 2003. (Photo: ESA)

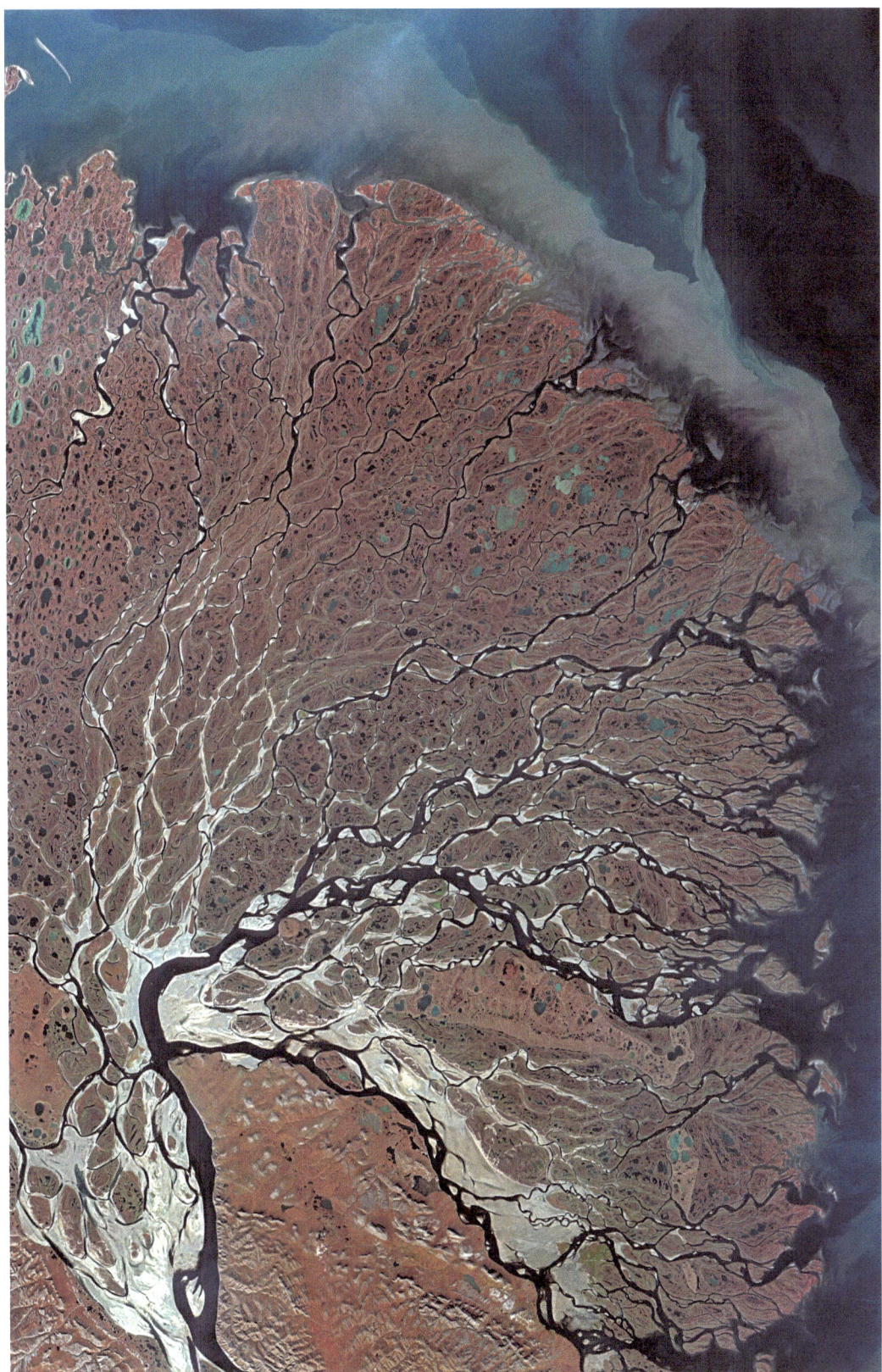

37 Siberia is crossed by numerous main rivers all of which drain into the Arctic Ocean. The longest of these is the 4400 km long Lena. The image from Copernicus Sentinel-2 presents its widely fanned delta, whose extent in north-south direction is up to 150 km and in east-west direction even more than 200 km. The many small lakes and ponds are formed as a result of the permafrost soil, which prevents the water from seeping. Copernicus Sentinel-2, September 9, 2020.

38 Water in frozen form constitutes the cryosphere of the Earth. It is found mainly at the two poles and in the inland ice of Greenland. A several kilometers thick ice sheet covers Antarctica almost completely. It has existed for about 15 million years and binds the largest part of the Earth's freshwater reservoir. At its edges, extensive areas of shelf ice have developed. The largest of these is the Ross Ice Shelf between Marie-Byrd-Land (right) and Victoria Land (left, with its ice-free dry valleys). Directly

in front of Ross Island with Mount Erebus, the southernmost volcano of the Earth, an almost 300 km long iceberg has detached from the edge of the ice shelf, which drifted along the coast in the following years. In the lower left part of the picture, the highest peaks of the Transantarctic Mountains protrude from the ice. MODIS/Terra, November 11, 2001. (Photo: NASA, GSFC, Jacques Descloitres, MODIS Land Rapid Response Team)

39 Opposite Ross Island, ice-free terrain can be seen in the upper picture. These are the dry valleys of Antarctica, located in Victoria Land. Due to the topography of the Transantarctic Mountains, the valleys are not exposed to the inland ice. This also leads to extremely dry winds that blow through the valleys and absorb residual moisture. Lakes constantly covered with an ice layer, under which there is

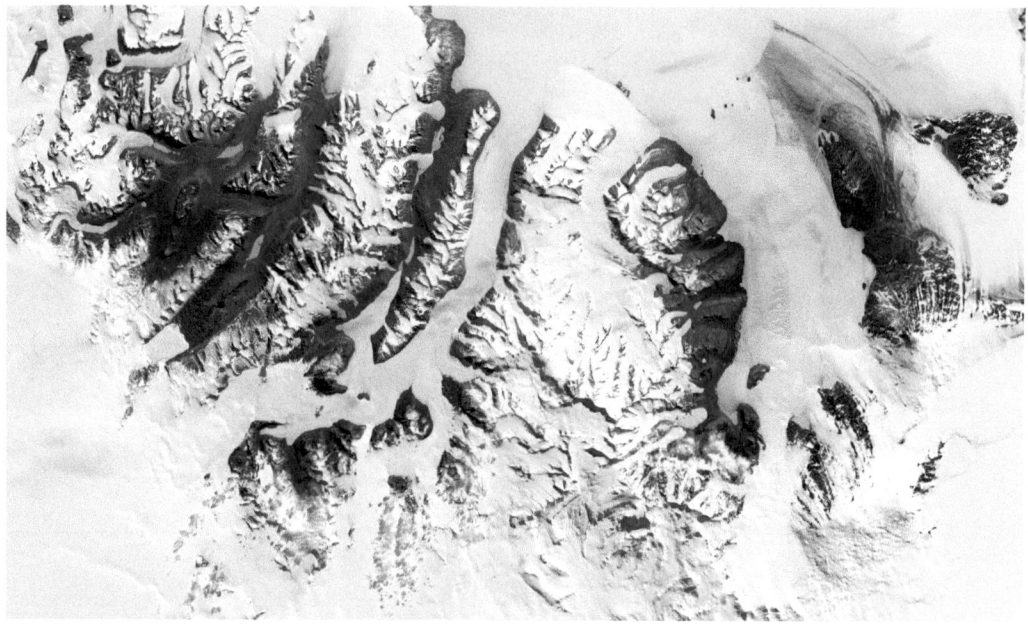

water in liquid form, and meltwater rivers that form in the Antarctic summer, allow the existence of microscopic life. Landsat-7, December 18, 1999. (Photo: NASA, Robert Simmon, Landsat-7 Science Team)

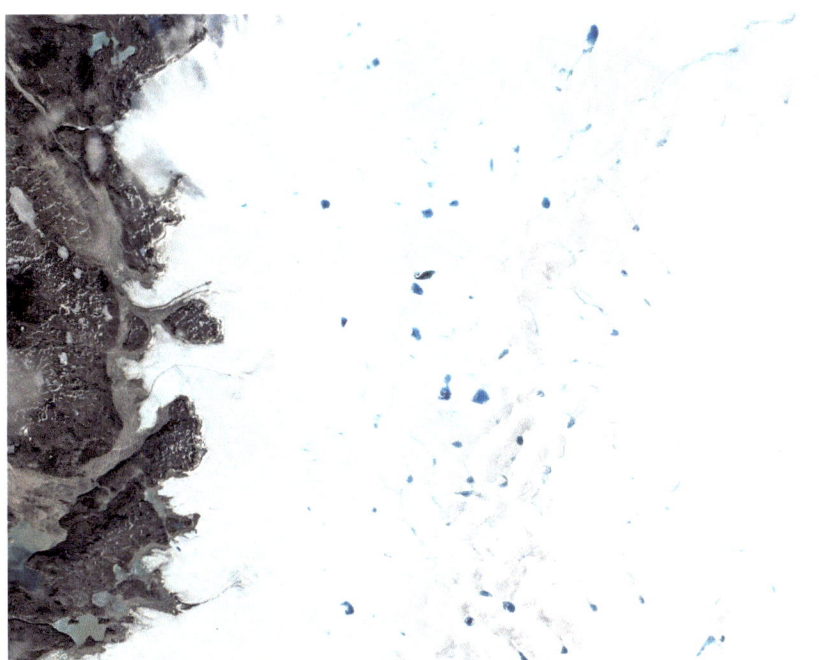

40 In the Arctic, the effects of climate change are clearly visible. Rising temperatures are putting a huge strain on the Greenland ice sheet. In this Landsat image from the beginning of summer in southwestern Greenland, a division in the color of the inland ice can be seen. On the far right, it is still covered with bright snow. When this melts, darker, old ice is revealed. On it, intensely blue colored meltwater ponds are becoming increasingly apparent. Through cracks in the ice, their water penetrates to the rock and acts there as a lubricant, which increases the flow speed of the glaciers. Dark ice also reflects less Sunlight. It absorbs more solar energy which in turn leads to higher melting rates. Landsat-8. June 21, 2013 (Photo: NASA, USGS, Jesse Allen, Robert Simmon, MODIS Land Rapid Response Team)

41 The characteristic triangular profile of Mount Everest, the highest mountain on Earth at 8848 m, is visible in the lower right half of the image. Diagonally opposite, the summit of the 8188 m high Cho Oyu rises. Numerous glaciers can be seen between the mountain giants. The Khumbu Glacier, which originates below the south flank of Mount Everest at an altitude of 8000 m, is the highest glacier on Earth. Even at these great heights, global warming is leading to an accelerated retreat of the glaciers. Copernicus- Sentinel-2, November 8, 2021.

42 Snow is the part of the cryosphere that also reaches more temperate geographical latitudes, as here in the eastern part of North America. A storm located about 1000 km southeast of New York brought heavy snowfalls, which reached as far south as West Virginia and as far north as Canada. New York and Long Island are also covered in snow. Cold air flows from the mainland towards the open Atlantic, where it forms "cloud streets" MODIS/ Terra, March 9, 2013. (Photo: NASA, Jeff Schmaltz, LANCE MODIS Rapid Response)

43 The image from the Suomi NPP satellite shows the northeastern USA with parts of Canada on a cold winter night. Chicago on Lake Michigan and New York with Long Island (right) are clearly visible, as is the reflecting snow cover on the ground. Suomi-NPP, January 8, 2015. (Photo: NOAA, NASA)

44 The Earth's atmosphere is only visible to an external observer when it contains solid particles, either condensed water droplets or dust and soot particles. Thus, a view from space is limited to the lowest atmospheric layers; where clouds form or where natural and anthropogenic phenomena introduce solid particles. Again, the already known "cloud streets" are visible. Cold air from the Greenland inland ice and the ice-covered polar sea flows west of Svalbard towards the open Greenland Sea, where these characteristic cloud structures form. MODIS/Terra, March 29, 2003. (Photo: NASA, GSFC, Jeff Schmaltz, MODIS Rapid Response Team)

45 Tropical cyclones are immediately noticeable by their massive cloud swirls, which rotate counterclockwise around the eye of the storm. Hurricanes originating in the North Atlantic move westward and often reach the East Coast of the United States with very high wind speeds. They occur most frequently there in September. The strongest hurricane of the 2003 season was Isabel. It reached the coast of North Carolina on September 18, 2003, causing severe destruction. MODIS/Aqua, September 18, 2003. (Photo: NASA, GSFC, Jeff Schmaltz, MODIS Rapid Response Team)

46 The MODIS image shows Kármán vortex streets, which are triggered by winds that hit the Canary Islands and Madeira. The islands act as an obstacle and lead to turbulence of the air flow. When clouds form in it, they lead to patterns named after the aviation pioneer Theodore von Kármán. MODIS/Terra, June 7, 2015. (Photo: NASA, GSFC, Jeff Schmaltz, MODIS Rapid Response Team)

 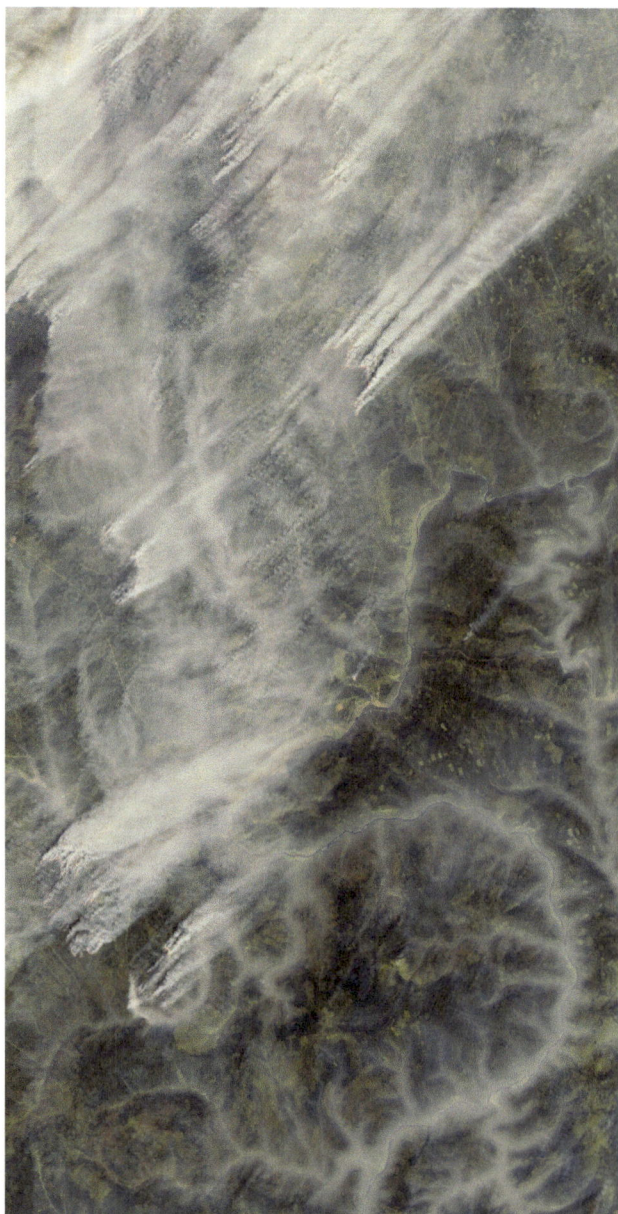

47 In the deserts of the world, extensive sandstorms can greatly impair the transparency of the atmosphere. Depending on the prevailing wind direction, desert sand is transported over long distances. Saharan dust even reaches Central Europe or the Caribbean in this way. In July 2020, extensive dust plumes spread from the Namib Desert across the Atlantic off the coast of Namibia, caused by hot dry winds from the interior of the country. The dust plumes extend from the Dorob National Park in the north to beyond Walvis Bay in the south. OLI/Landsat 8, July 17, 2020. (Photo: NASA, Lauren Dauphin, USGS)

48 Smoke from vegetation fires contributes significantly to local air pollution. In the spring of 2016, there were extensive forest fires in Siberia, triggered by extremely dry and warm weather. In the administrative district of Irkutsk large areas of the boreal forest burned. The smoke from the fires formed dense northeastward moving plumes or filled the river valleys of the region like that of the meandering Angara. The fire sites can be recognized by their reddish-yellow color at the origin of the smoke plumes. Copernicus Sentinel-2, September 28, 2016. (Photo: contains modified Copernicus Sentinel data (2016), processed by ESA)

49 In some areas of the world, a severely limited view is immediately noticeable when looking from above. Unlike the mostly bright cloud areas, such areas appear pale and gray. Obviously, this is polluted air. Often densely populated regions with high industrial use are hidden under such smog domes. The situation can be particularly critical in China as a country with high growth in this respect. The combustion of coal still plays a major role industrially and in households. When high pollutant emissions are then enriched by the weather, the situation becomes dramatic. This was the case in eastern China in early October 2010. In Beijing and the surrounding area, visibility was only 100 m, even the Sun was only weakly visible at noon. MODIS/Terra, October 8, 2010. (Photo: NASA, Jesse Allen, MODIS)

50 Not all clouds are of natural origin. Contrails are created in aviation when hot engine exhaust gases are emitted into the cold atmosphere of today's flight altitudes. In areas with heavy air traffic, like here over the Channel between France and England, contrails, which slowly dissolve and stay in the sky for several hours, can indeed lead to a change in cloud cover. MODIS/Aqua, December 9, 2003. (Photo: NASA, GSFC, Jacques Descloitres, MODIS Rapid Response Team)

51 Left: Earth is the only known planet with a biosphere. Chlorophyll is one of the substances that determine the color appearance of our planet. It gives plants and thus the vegetation zones on Earth their green color. In temperate and high latitudes, leaf plants reduce their metabolism in autumn and the leaf green is broken down. Yellow and red pigments come to the fore. In the north of America, the forests turn red, it is "Indian Summer". Exactly this phase prevails here around Lake Superior with intense coloring of the forest areas. MODIS/Terra, October 11, 2004. (Photo: NASA, GSFC, Jacques Descloitres, MODIS Rapid Response Team)

52 Below: The Earth's biosphere also includes every single-celled organism. As soon as these simple individuals appear in masses, they can even be observed from space. When algae and phytoplankton rapidly multiply in waters, it leads to algae bloom. The chlorophyll of the microscopic beings then colors large areas of the water. In the summer of 2011, an extensive phytoplankton bloom was observed in the Barents Sea north of the North Cape. MODIS/Terra, August 14, 2011. (Photo: NASA, GSFC, Jeff Schmaltz, MODIS Rapid Response Team)

53 The Amazon rainforest extends almost across the entire width of northern South America. It houses a biosphere with enormous biodiversity. At Manaus, visible in the left half of the image slightly below the center, the wide Rio Negro with its dark water meets the main course of the Amazon. Only after several kilometers does the water of the Rio Negro fully mix with the light water of the Amazon. MODIS/Terra, September 8, 2002. (Photo: NASA, GSFC, Jacques Descloitres, MODIS Rapid Response Team)

54 Below: In agriculture, chlorophyll forms abstract patterns. To use the available groundwater as sparingly as possible, rotating irrigation systems are used here in Kansas. This results in circular agricultural cultivation areas for grain, corn, and millet. Their different coloring indicates different stages of growth. Copernicus Sentinel-2, June 29, 2020.

55 The world's population increasingly lives in large cities. Today's satellite images allow very detailed insights into urban areas. Here you can see Munich, taken with a sensor, that works in the meter range. The bare Theresienwiese stands out clearly to the left of the center of the image. IKONOS, March 22, 2003. (Photo: European Space Imaging)

56 Satellite images of the last decades show how urban centers emerge and often grow uncontrollably. One of the early highly developed metropolises is Rome, the Eternal City. During the time of the Roman Empire, it already reached the status of a million city. Today, 2.7 million inhabitants live in its urban area on the Tiber and the surrounding hills. Copernicus Sentinel-2, September 17, 2020.

57 A synonym for a modern metropolis was for a long time New York. It was considered the largest metropolis, but today with a population of just over 8 million, it ranks far behind numerous other cities. Manhattan with the rectangular-looking Central Park and its skyscrapers, although now overshadowed by taller buildings in other cities, has nevertheless lost none of its charm. Copernicus Sentinel-2, September 19, 2020.

58 A cradle of space travel and thus the basis for us to be able to view all the images compiled here at all, are the Cape Canaveral Space Force Station of the US Air Force (right on the coast) and the John F. Kennedy Space Center (above the center of the image) in Florida. Already used by the American Air Force for rocket launches in the 50s, it was from the 60s onwards the place where all manned US space flights began and from which numerous unmanned satellites took off into space. The lagoons of the Banana River and the Indian River are colored green by strong algae bloom. Copernicus Sentinel-2, December 9, 2020.

59 Impact structures bear witness to Earth's collisions with small objects of the solar system. Like all other members of the solar system with a solid surface, they are also found on Earth, but not in large numbers. This is due to the strong terrestrial tectonic activity as well as erosion, weathering or sedimentation. Over time, they obscure a formed crater beyond recognition or hide it underground. The Meteor Crater in Arizona, with a diameter of 1.2 km, was created by the impact of an approximately 50 m measuring iron asteroid about 49000 years ago on the plain of the southern Colorado Plateau. One can still find numerous iron meteorites in its vicinity, named after the Canyon Diablo, recognizable to the left of the crater. Copernicus Sentinel-2, September 27, 2018.

60 Much older than the Meteor Crater in Arizona is the 30 km large Shoemaker impact structure in Australia, named after the geologist Shoemaker, visible on the right half of the image. It was formed between about 600 million and 1.3 billion years ago. Today it is particularly noticeable due to the multitude of shallow, ephemeral salt lagoons that periodically form within it. Their varying depth and sediments dissolved in the water create colorful patterns within it. Copernicus Sentinel-2, July 3, 2017.

61 A night shot of North America with the brightly lit cities from the East Coast to the Midwest (above). On October 4-5, increased activity was registered on the Sun. Three days later, this stream of particles arrived at Earth and triggered strong auroras. The day/night sensor of Suomi-NPP documented the bright band of an Aurora Borealis over Canada. The aurora visible here in the grayscale image actually has a greenish color. Suomi-NPP, October 8, 2012.

There are also auroras in the south, but not as Aurora Borealis, the "Northern Lights" but as Aurora Australis, the "Southern Lights" (below). Artificial lighting like in the above image is missing in the uninhabited Antarctica. The Aurora Australis is shown here over the Antarctic Queen Maud Land. Suomi-NPP, July 15, 2012. (Photos: NASA, NOAA, DoD, University of Wisconsin, Jesse Allen, Robert Simmon)

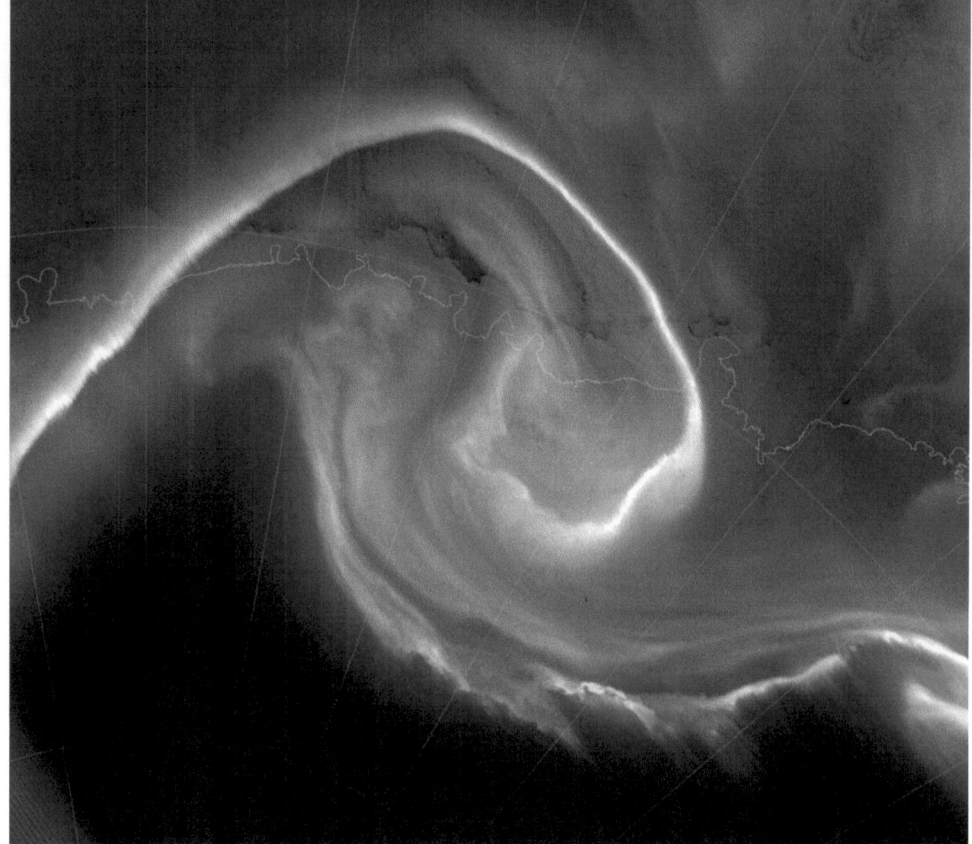

62 Global Earth views, composed of many individual images from a single day were in the early days of Earth observation with satellites a difficult undertaking. For the mosaic of global cloud cover, 480 TIROS-9 individual images taken in 24 hours had to be assembled in the mid-1960s (above). Since the images have a temporal offset, some cloud fields appear twice at the left and right edge. Today, in the digital age, a computer performs the tasks of assembling individual images into a mosaic. From the numerous individual images covering the entire Earth, the global cloud cover on an Earth without night was created. TIROS-9, February 13, 1965 and MODIS/Terra, July 11, 2005. (Photos: NOAA - above, NASA - below)

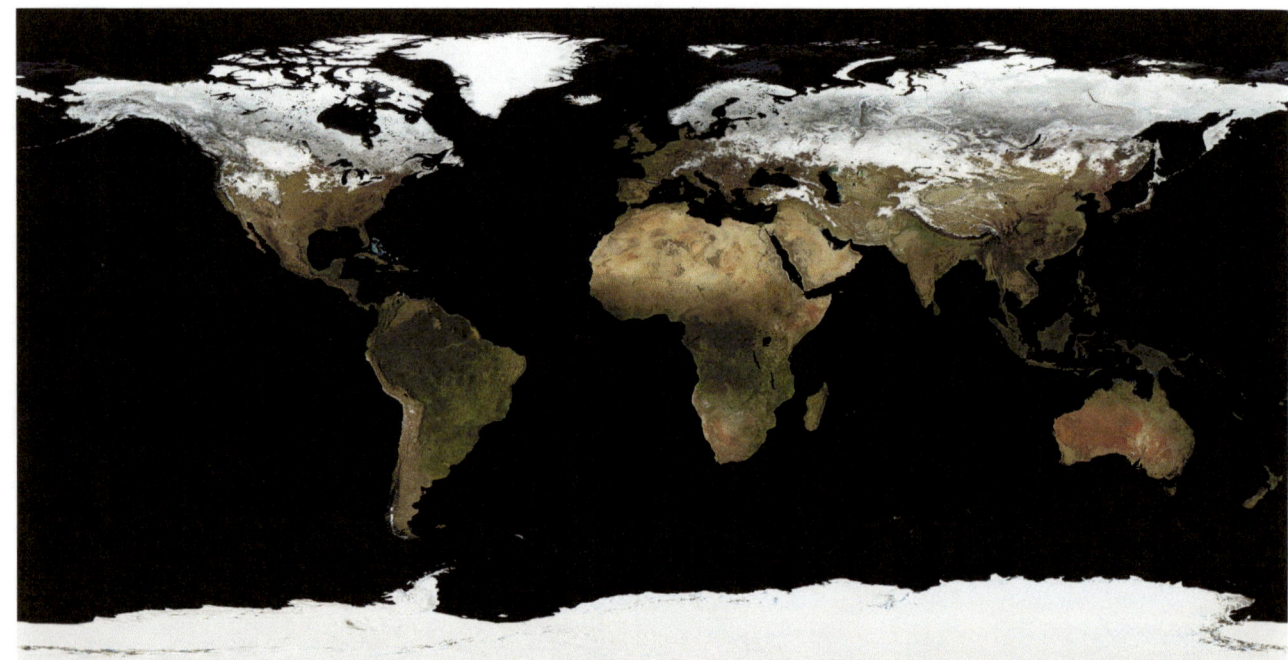

63 Digitally created virtual views from low Earth orbit are excellent for visualizing seasonal fluctuations on Earth, for example the coming and going of leaf green as a sign of changing vegetation and the variable coverage with snow and ice. The changes between winter (January 1-31, 2004, above) and summer (July 1-31, 2004, below) is particularly apparent. MODIS/ Terra. (Photo: NASA, Reto Stöckli)

64 The individual images of a satellite in low Earth orbit can be projected arbitrarily in the computer, also onto a sphere. Then you get a spatial image of the Earth, over which you can position yourself as you wish. From MODIS data from February 2002, a view of the western and eastern hemisphere of the Earth was generated from a distance beyond low Earth orbit. It resembles the images of geostationary satellites. MODIS/Terra, February 2002. (Photo: NASA, GSFC, Reto Stöckli, Robert Simmon, MODIS Land Group, MODIS Science Data Support Team, MODIS Atmosphere Group, MODIS Ocean Group, USGS EROS Data Center, USGS Terrestrial Remote Sensing Flagstaff Field Center, Defense Meteorological Satellite Program)

65 Although the majority of Earth observation satellites move in polar orbits, there are few overall images of the vast regions of the Arctic and Antarctic. These either come from Apollo astronauts after they had set their course towards the Moon, from interplanetary probes performing a swing-by maneuver at Earth, or more recently from satellites in a Molniya orbit. Usually, views of the Earth's poles can only be created from photos of low Earth satellites by combining the images of several overflights in one day. The mosaic then generated in the computer with a polar projection directly from above spans 360° in geographical longitude. Only the constantly changing cloud cover sometimes prevents cloud fields from being combined completely congruently.

The Arctic is shown in the upper right after the start of summer. The ice cap of Greenland appears bright white in contrast to the North Polar Sea covered with darker ice to its right. Iceland, Norway, and Sweden, as well as Finland, can be seen in gaps in the clouds. The landmass of Russia stretches along the right half of the picture. In Siberia, extensive gray smoke plumes caused by forest fires can be seen.

Antarctica, also at the beginning of summer in the southern hemisphere, appears quite different in the lower picture. Completely surrounded by oceans, it is only inhabited by the staff of research stations. At the northern edge of the picture, you can see the southern tip of South America, with the O'Higgins Peninsula of Antarctica stretching towards it. Only the dry valleys in Victoria Land and peaks of the Transantarctic Mountains, which adjoin it in an arc, reveal some dark-looking ice-free spots. MODIS/Terra, July 3, 2021, and January 7, 2021. (Photo NASA)

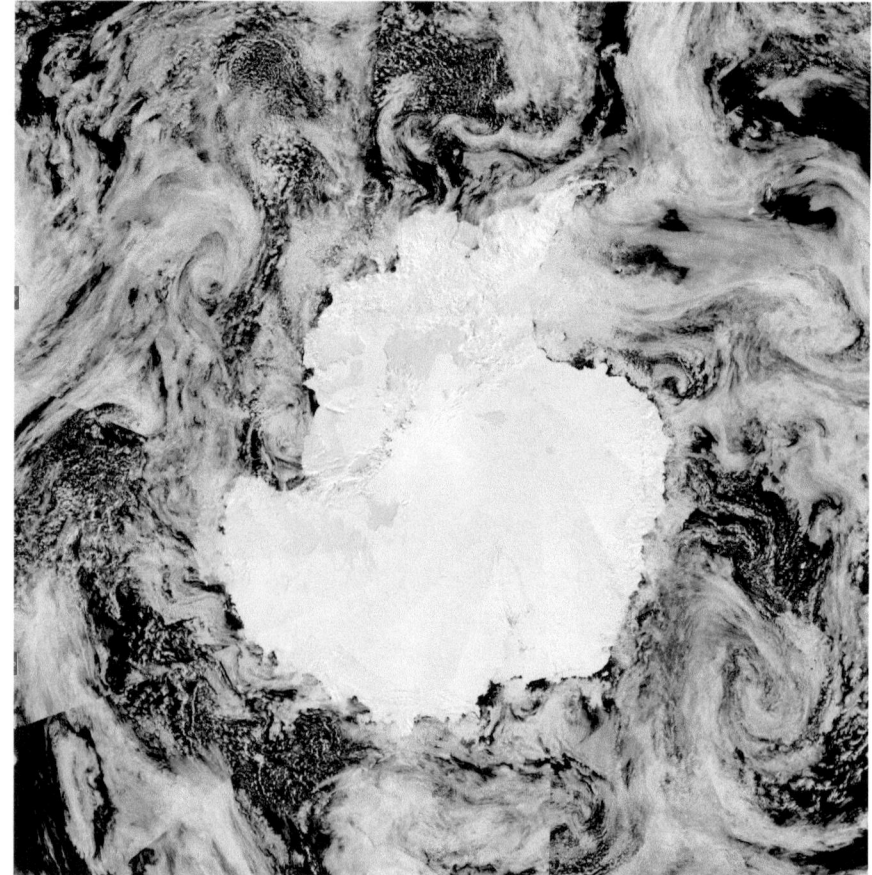

66 If you combine individual night shots into an overall image of the Earth, virtual views of our home planet immersed in darkness emerge, illuminated only by the bright glow of the cities and the rising Sun at the edge of the Earth. From over half a year of collected images from Suomi-NPP, nocturnal hemispheres were created. On the left, you can see Europe and Africa with parts of Asia. In Europe, the Benelux countries stand out together with the Ruhr area, as well as the Po Valley in Northern Italy. Africa, on the other hand, is largely immersed in darkness. In Egypt, the Nile Delta is clearly visible. Below is the American hemisphere, with densely populated North America and its numerous large cities standing out, while metropolises in South America are mainly found in coastal regions (left). The lower right image shows Asia and Australia. Here, the transition from bright North India with the Ganges plain to the dark Tibetan highlands catches the eye. Suomi-NPP, April-October 2012. (Photo: NASA, NOAA, Robert Simmon)

5.   Manned Low Earth Orbit - Beautiful Views

Earth views taken by machines are highly interesting, but they do not fully satisfy human curiosity. There is always the question of how a human sees the Earth from space and what his impressions are. Answering these questions would certainly not have been motivation enough to send an astronaut or cosmonaut into space in small capsules and bring them back safely to Earth. The desire to show that one's own system is better than the supposed opponent's fueled the pursuit of "always higher, always further" during the Cold War. A technical optimism, sometimes appearing somewhat naive in retrospect, led to a belief in the conquest of space, which made the presence of humans outside the Earth's atmosphere seem necessary. The announcement by John F. Kennedy in 1961 of a successful Moon landing of an American by the end of the decade had officially opened the race to the Moon and made manned space travel a focal point in the space program on both the American and Russian (Soviet) side. Initially, in the pioneering days of space travel when attempts were made to penetrate space with simple rockets, the goal of a manned orbit above the Earth's atmosphere was not necessarily on the agenda. The interest was more generally in how to transport people to great heights and how the pilot's reactions to the occurring stresses were. When V2 rockets launched from White Sands reached speeds of several 1000 kilometers per hour, no human had yet traveled at simple sound speed. On the other hand, aeronautical achievements provided an excellent platform to demonstrate the technical capabilities of a nation. In the United States, therefore, there was great interest in the rocket plane programs. The first, the Bell X-1, accelerated Chuck Yaeger on its 50th flight in October 1947 for the first time to more than 1000 kilometers per hour; the sound barrier was overcome. Another representative of the X-program, and probably one of its highlights, we had already met in the third chapter with the rocket plane X-15, which could briefly overcome the boundary to space. These aeronautical projects gathered experiences on how humans react at great heights and under enormous accelerations; however, they did not lead to sending someone into low Earth orbits for longer periods of time. Pilots had to become astronauts. A precursor of the American space agency NASA, the NACA (National Advisory Committee for Aeronautics) founded as early as 1915 dealt with the challenges of manned movement at high speeds and at great heights from the mid-40s.

Initially other living beings had to show that such journeys could be survived unscathed. Monkeys and mice were the test animals on the American side; mice, rats, rabbits, and dogs on the Russian side, where similar goals existed, but always shielded from the public eye. The animals initially paid tribute to the rudimentary technology and did not survive their missions. Only in 1951 did an animal crew of monkeys and mice return safely from 73 kilometers height, two Russian dogs survived their suborbital excursion in the same year. All these activities were only precursors to the ambitious goal of sending a human around the Earth. In 1958, the Mercury program was officially announced to enable the first space flight of an American in a space capsule. The astronaut team consisted of seven specially selected pilots from the US military. In their public perception, they were initially below the test pilots used in extreme flight experiments. Since they could not control their space capsule themselves, they were essentially only considered as publicity-effective ballast.

Before, following a series of ballistic launches, John Glenn orbited the Earth in February 1962 in the space capsule Friendship 7 on the third Mercury flight three times, the Russian side had already sent Yuri Gagarin into space as the first cosmonaut in April 1961. Although since then only China has developed the ability to send people into space and bring them back with their own means, representatives of numerous nations were given the opportunity to view our home planet from near-Earth orbits. Joint projects created regularly perceived co-flight opportunities since the 70s. They culminated in the International Space Station

M. Gottwald, *The Earth*,
https://doi.org/10.1007/978-3-662-69633-0_5

as a global joint venture. During the Mercury era, however, this was still far off. People dreamed of the outstanding goal of a manned Moon landing and could hardly imagine how this enthusiasm could ever wane again. After four successful flights, the Mercury program ended in May 1963 and was followed by flights of the Gemini capsules with two men. It took place between 1965 and 1966. The Gemini missions were already significantly oriented towards the requirements of future Moon flights; astronauts could actively intervene in the flight events and perform necessary orbital maneuvers. The Gemini program was followed by the series of Apollo Moon flights. Two of the Apollo missions were limited to near-Earth orbit - the first flight Apollo 7 in October 1968 and Apollo 9 in February 1969 for testing the lunar module.

The USSR also established a manned space program after Gagarin's flight with increasingly complex space capsules. Their intentions regarding a manned visit to our lunar companion were then hidden, but existing plans were abandoned after the first successful Moon landing of Apollo 11 and they focused on the development of space stations in near-Earth orbit, the first launch of which was Salyut-1 in April 1971.

The ballistic Mercury flights, which initially transported chimpanzees briefly into space, had automatic cameras on board for documenting the ascent and descent. They recorded in good quality what could be seen from the window of the space capsule. When astronauts were transported, observations of the Earth were not at the top of the list of tasks to be accomplished. Initially, the main focus was on understanding how humans function and react in zero gravity. It is therefore not surprising if the photography of the first two Mercury missions with crew initially yielded unsatisfactory results. John Glenn carried a simple modified 35mm camera. Its handling with a spacesuit in the narrow capsule was problematic. Only the last two Mercury astronauts Walter Schirra and Gordon Cooper received any brief introduction to the possibilities of Earth photography from space. They then also had a modified Hasselblad medium format camera, whose larger negative format allowed for higher quality images. Cooper was also able to invest much more time in photographing our planet on his 34-hour flight. What the last two Mercury missions hinted at, continued with the Gemini

ventures. With the Hasselblad as workhorse, they achieved extraordinarily impressive Earth images for that time. Its square 60 mm film format became the standard for photography on manned NASA missions. While the unmanned probes still transmitted relatively inconspicuous images in black and white, the astronauts now brought back sharply focused photos from their excursions - and all in color!

Of the two Apollo flights, Apollo 9 is particularly noteworthy, as the arsenal of cameras carried not only documented the testing of the lunar module and the flight maneuvers, but also for the first time multispectral images of the Earth's surface were created. For this, the same scene was photographed with four identical cameras through different filters. For practical Earth observation these images were only of experimental character. Only unmanned probes have later made the multispectral approach successful, where the medium film never played a role.

Do the Earth views taken by astronauts differ from those obtained with automatic probes in near-Earth orbit? In the early days of space travel, certainly in terms of their quality, as a correctly exposed color film with a Hasselblad definitely delivered better images than a line-by-line scanned black-and-white TV image. Also, the size of the depicted scene may differ, as manned near-Earth missions move at altitudes between 250-300 kilometers, while many Earth observation satellites orbit at altitudes up to 800 kilometers around the Earth. With the same optics and geometry, therefore, better geometric resolution is achieved from manned space capsules. However, unmanned platforms on polar orbits fly over far north and south regions of the Earth, which can never be reached by manned missions due to the lower orbital inclination. Finally, images of Earth are often taken from an oblique perspective by astronauts. In contrast, an Earth observation satellite must represent the Earth's surface as accurately as possible. This is achieved by having the imaging instrument look directly down at the Earth. Only very few special tasks require the alignment, for example, towards the Earth's horizon. Astronauts, on the other hand, can look down and sideways out of the spacecraft. Depending on the subject, they then capture what appears attractive. This is often a wide view from 250-300 kilometers high onto the distant curved

Earth's horizon. Many such images are only intended to show the beauty of our home planet to an interested public that has remained on Earth. Neither the Mercury, Gemini or Apollo astronauts had ample time for extensive photo series. The flight requirements were simply defined differently; also the tightness in the space capsules - initially even intensified by constantly worn spacesuits - was a hindering side effect.

When in December 1972 after Apollo 17 the manned Moon flights ended were, manned space travel in the USA needed new goals. A relatively obvious one was to use existing rocket parts from the Moon program to build a space station. This space laboratory launched in May 1973 under the name Skylab was visited by three crews until February 1974. More than five years later, in July 1979, the empty Skylab laboratory had descended so far in its orbit that it entered the Earth's atmosphere and crashed onto Earth. Due to its size, some parts even remained intact and fell onto uninhabited terrain in Australia. The photographic equipment of the Skylab crews was very extensive. In addition to 35mm, medium format, film and TV cameras, there was also a specially designed experiment for Earth observation, with which multispectral recordings were possible. Exactly at the time when the crews of Skylab conducted their experiments, however, Landsat-1 was already the first representative of modern automatic Earth observation satellites with multispectral sensors in orbit. Anyone hoping for further views of Earth from manned spacecraft after Skylab had to wait patiently until 1981.

Although in 1975 in the meantime the joint flight of Apollo-Soyuz took place as the first American-Russian cooperation in space, but here the political aspect was in the foreground and the undertaking was primarily of symbolic character. Also from the Soyuz capsules launched again and again from Baikonur on the other side of the globe, no exciting views of Earth were made accessible to a wide public. With the first flight of the Space Shuttle Columbia in April 1981 from Cape Canaveral in Florida then began a three-decade phase of new views of the Earth. The expectations for the Shuttle program were high, as it promised cost-effective and rapid access to the low Earth orbit with possibilities ranging up to satellite repairs and satellite launches virtually "in flight" Today we know that these goals were much too ambitious. The Space Shuttles turned out to be ultimately too expensive. They have since been retired. What they certainly attracted enormous attention within in the last 30 years of their use, are the countless photos of our home planet; not necessarily suitable for Earth science studies, but excellent documents that show the beauty and uniqueness of the Earth. Shuttle crews always carried a comprehensive set of recording systems. These ranged from 35mm cameras to the Hasselblad medium format cameras already known from Gemini and Apollo flights, up to a large format device for recordings with a magazine in the format 23 cm × 46 cm. Such a monster was no longer suitable for handheld photography. It was housed in the cargo bay and had to be operated like a scientific instrument.

Even better than during shuttle missions, astronauts can view the Earth from space stations. With permanent occupancy, there is enough time besides scheduled tasks to take aim at the Earth. The precursor to the International Space Station ISS, the Russian MIR station, made its debut from 1986 until its controlled crash in 2001. It paved the way on political terrain for intensive cooperation between space agencies of the eastern and western hemispheres and allowed, also through longer visits of the Space Shuttle, to better understand the technical conditions before construction of the ISS began in 1998. MIR occasionally had a docked module equipped with instruments for observing the Earth. The devices housed there could certainly compete in their performance with those on automatic platforms. They also had the advantage of more precise detail resolution due to the lower orbit. But ever since Skylab, the peculiarities of manned missions have prevented a meaningful continuous use in the study of the Earth from space. No manned mission moves on a polar orbit. These restrictions of the orbit make it impossible to fly over the entire Earth. If one also considers the effort that had to be made in manned missions then and now to ensure the safety of the crew, it is not surprising that the continuous exploration of the Earth from space demanded automatons, which ultimately proved to be the more suitable way. Recordings of unmanned missions are calibrated and can be brought into any cartographic projections with suitable algorithms. They appear almost perfect even though the image is often only a by-product generated from the scientific data. Not so with the images obtained by astronauts. They often have the aura of a spontaneously taken travel photo. One immediately recognizes which motifs appear particularly frequently and therefore look particularly attractive from orbit. Over continents, these are mainly untouched, desolate landscapes such as desert areas and mountains, in addition to cities, rivers, and lakes. oceans are particularly interesting where land and water meet, i.e., coastal regions or the numerous island worlds in the vastness of the world's oceans. Due to the usual orbit inclinations of manned missions, very high northern and southern latitudes are usually not flown over, which is why distant images of the polar regions only appear very rarely. Many of the photos taken at a shallow angle, especially if they are shot with telephoto lenses, then they look as if they were produced from a slightly too high-flying airplane. A genre of photography from the ISS is now the topic "Earth at Night" which shows us a variety of regions when the space station is on the dark side of the orbit. These images appear different than, for example, the night scenes of the automatic Suomi-NPP, which we got to know in the illustrations of the previous chapter, simply because of their greater proximity to the Earth's surface and their color. The possibilities for spontaneous photos on the International Space Station ISS, which had reached its planned final expansion phase around 2010, are unsurpassed. This includes a module specifically attached for observing the vicinity of the space station, whose seven windows provide a view of the immediate surroundings of the station and the Earth. This "Cupola"; facility is not only useful for external work that is controlled from inside the ISS; it offers astronauts, who are on a long-term stay, a permanent connection to their home, the psychological effect of which should not be underestimated. According to current plans for the operation of the ISS, its crews will continue to provide us with images of the Earth for several more years. When selecting suitable images for this stage, we face the same dilemma as with unmanned ventures in low Earth orbit. The sheer amount - the number of Earth views taken on the ISS has now exceeded the million mark - makes it difficult to get an overview. Here too, after some historical excursions into the 60s and 70s, we follow the scheme of lithosphere, hydrosphere, cryosphere, atmosphere, and biosphere with the anthroposphere as a by-product. Just like with unmanned missions, images have proven to be suitable for our purpose, showing that the Earth is part of the solar system, such as glimpses to the Earth's horizon with the night sky glow (airglow). Auroras as an example of the interaction between the Sun and Earth. Standing above the dense atmosphere, the Moon is a frequent subject and much rarer, but also more spectacular, sometimes you see a comet on the horizon. In shots that look like astrophotos from the Earth's surface, you finally get an impression that we are indeed just a part of the universe.

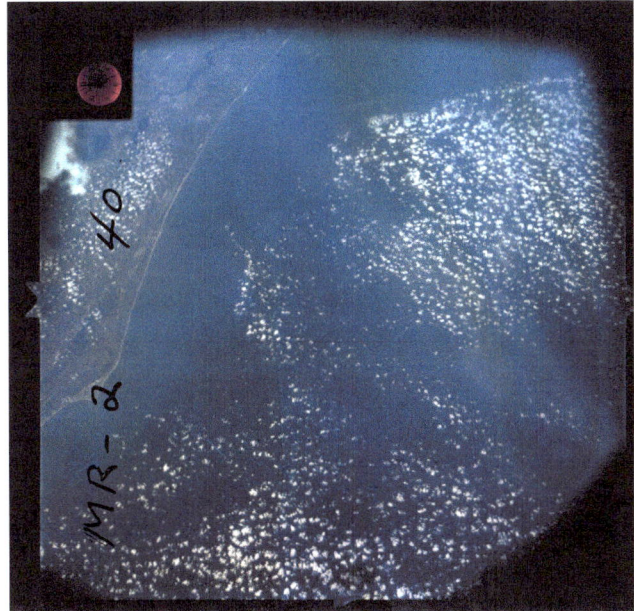

67 That was the view for the chimpanzee named Ham, when he was launched 250 km into space by a Redstone rocket on a ballistic flight in January 1961. An automatic camera documented the ascent and descent. You can see the coastline of Florida with the launch site Cape Canaveral. Mercury-Redstone 2, January 31, 1961. (Photo: NASA)

68 In September 1961, for the first time, a Mercury capsule occupied with a simulator instead of an astronaut was sent on a successful test flight around the Earth. As with the previous tests, a camera automatically exposed a film. This shot was taken while flying over the west coast of Africa at about the height of the Canary Islands with the Anti-Atlas stretching northeast. Mercury-Atlas 4, September 13, 1961. (Photo: NASA)

69 John Glenn orbited the Earth as the first American in February 1962. One of his shots was taken approximately at the same place as the previous one of the Mercury test flight but a little further inland and looking west. Glenn carried for Earth shots only a converted, difficult to operate 35mm camera with him while in the unmanned earlier ventures automatic cameras had been used. The difference in quality is obvious. Mercury-Atlas 6, February 20, 1962. (Photo: NASA)

70 On the last two Mercury missions medium format cameras took over the photographic work. The image quality benefited significantly from this. Here is a shot of Gordon Cooper's last Mercury flight with the highlands of Tibet and its numerous lakes. Mercury-Atlas 9, May 16, 1963 (Photo: NASA)

71 The dry northern part of Baja California with the mouth of the Colorado River appears almost surreal in this shot from Gemini 4 because of its coloration. Gemini 4, June 5, 1965. (Photo: NASA Johnson Space Center, GEM04-08-34672)

72 A view of the east coast of Florida with the rocket launch site Cape Canaveral. Gemini 5, August 28, 1965. (Photo: NASA Johnson Space Center, GEM05-02-4559)

73 The astronauts of Gemini 9 looked here at the 3800 m high Lake Titicaca and a part of the Bolivian and Chilean Andes as well as the extremely dry Atacama desert and its salt lagoons on the left edge of the picture. Gemini 9, June 5, 1966. (Photo: NASA Johnson Space Center, GEM09-C-38312)

74 The Canary Islands off the west coast of Africa. In the front La Palma, followed by El Hierro, Gomera, the main island Tenerife, Gran Canaria and Fuerteventura. Lanzarote is hidden in the north under clouds. Gemini 9, June 6, 1966. (Photo: NASA Johnson Space Center, GEM09-B-38441)

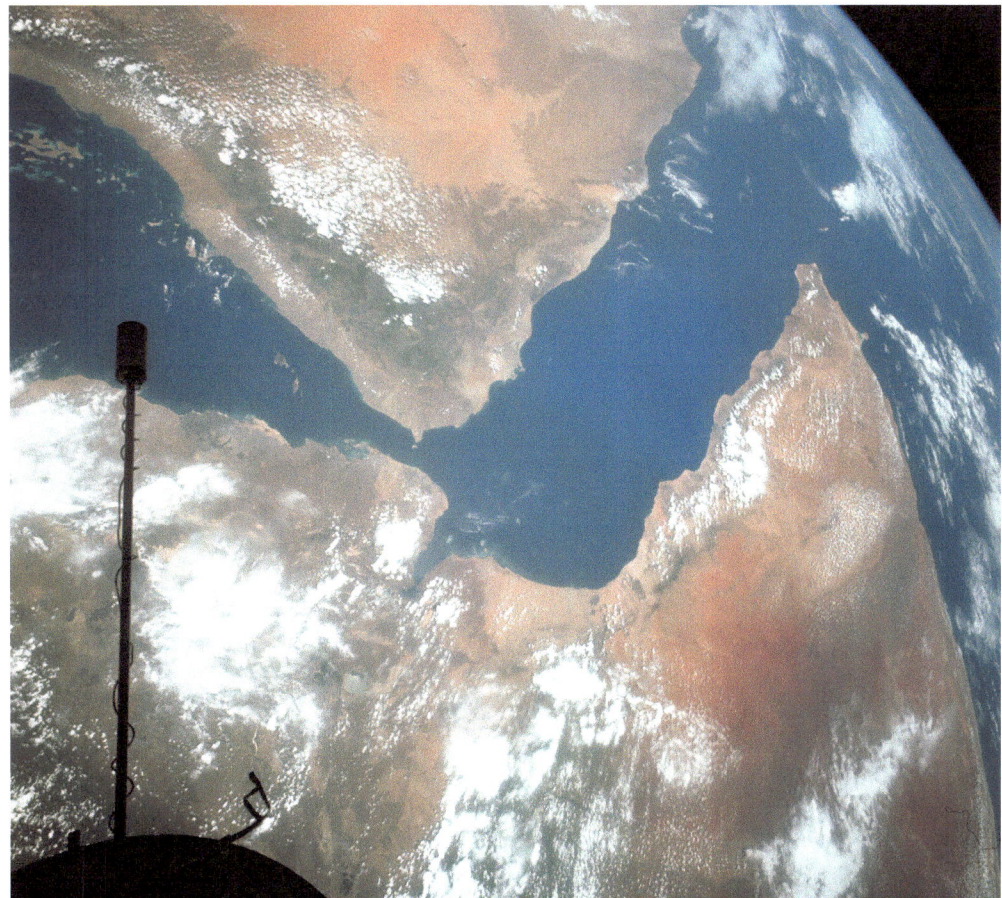

75 The Horn of Africa and on the other side of the Gulf of Aden the southwest corner of the Arabian Peninsula. Gemini 11, September 14, 1966. (Photo: NASA Johnson Space Center, GEM11-9-54536)

76 Gemini 11 reached the greatest with 1370 km Altitude of all manned space flights in low Earth orbit. When this image of the Indian subcontinent was taken, the spacecraft was at an altitude of 400 km. The view extends far to the north. On the horizon, beyond the Ganges plain, the rising Himalayas can even be faintly seen. Gemini 11, September 14, 1966. (Photo: NASA Johnson Space Center, GEM11-8-54676)

77 In the 1960s, the astronauts of the first Apollo mission saw Lake Chad as extensive lake. Today, the water level has dropped massively and the lake only has a fraction of its former water surface. Apollo 7, October 11-22, 1968. (Photo: Earth Science and Remote Sensing Unit, NASA Johnson Space Center, AS07-8-1932)

78 Apollo 9 was the last Apollo mission in Earth orbit. In early March 1969, the high plateaus of Arizona were covered with a thin layer of snow. The long incision of the Grand Canyon stands out from it. Apollo 9, March 12, 1969. (Photo: Earth Science and Remote Sensing Unit, NASA Johnson Space Center, AS09-20-3137)

79 One of the cameras that belonged to Skylab's Earth observation experiment was able to capture small sections of the Earth's surface with high resolution on film. Its results resembled those of multispectral sensors on unmanned satellites. This is evident in the image shown here of the Colorado and Green Rivers in the Canyonlands southwest of Moab. Skylab 2, May 26 - June 22, 1973. (Photo: Earth Science and Remote Sensing Unit, NASA Johnson Space Center, SL2-81-016)

80 The last of the Skylab crews captured New Zealand in the picture. To the right is the North Island, to the left the South Island, and in between is the Cook Strait. The grid markings in the image helped calibrate the shots. Skylab 4, November 16, 1973 - February 8, 1974. (Photo: Earth Science and Remote Sensing Unit, NASA Johnson Space Center, SL4-142-4592)

81 The orbits of manned space missions in low Earth orbit are all similar. Depending on the inclination, the spacecraft reach maximum moderate northern and southern latitudes. Thus, they always fly over the same areas. This is also reflected in the photos. One region that has always prompted astronauts to press the shutter since the 1960s is the Middle East, especially the Sinai Peninsula with adjacent Israel. In March 2002, the crew of the 108th Space Shuttle flight captured this view of Sinai, Israel with the Jordan Valley and the Dead Sea, and to the east of it, Jordan and Syria. This flight of the Space Shuttle (STS) was one of the last missions that still used analog film material. STS109, March 2, 2002. (Photo: Earth Science and Remote Sensing Unit, NASA Johnson Space Center, STS109-708-024)

82 With the observation platform "Cupola" installed in 2010, the astronauts on the ISS not only have the opportunity to monitor activities outside the space station, but can also use it for observations of the Earth. One of the first photos from the Cupola shows the view from its seven windows onto the Sahara below. ISS, February 17, 2010. (Photo: Earth Science and Remote Sensing Unit, NASA Johnson Space Center, ISS022-E-066972)

83 Actually, the ISS is looking at a relatively unspectacular scene in North America here. In fact, this photo can be seen as a reminiscence of the Explorer II balloon expedition from the year 1935 (Chapter 2). Slightly to the left of the center of the image, one can recognize the dark oval of the Black Hills, from whose Stratobowl Explorer II rose. Afterwards, the balloon drifted over the plains of South Dakota and reached there, in the ISS image to the right of the Black Hills, the then record height of 22 km. ISS, October 20, 2014. (Photo: Earth Science and Remote Sensing Unit, NASA Johnson Space Center, ISS041-E-84308)

85 The Alps were formed as part of the Alpine orogeny, whose fold mountain chain, including the Himalayas, extends from North Africa to Southeast Asia. Here in the distance, one can see the Lake Geneva and further east Lake Constance. ISS, October 13, 2018. (Photo: Earth Science and Remote Sensing Unit, NASA Johnson Space Center, ISS057-E-051220)

84 A panorama of the Himalayas, composed of six shots, as the International Space Station moved at altitude of 360 km over the Tibetan highlands. The panorama looks south and covers a length of more than 130 km. The two highest mountains in the right part of the image are Makalu at 8462 m (left) and Mount Everest at 8850 m (to the right of it, with a long cloud trail in the background). ISS, January 28, 2004. (Photo: Earth Science and Remote Sensing Unit, NASA Johnson Space Center, ISS008-E-13302 to 13307)

86 The massif of Kilimanjaro is located on the eastern edge of the East African Rift. It is of volcanic origin and was formed there about 2-3 million years ago during the breaking apart of the African Plate. Today, Kilimanjaro is considered a dormant stratovolcano, consisting of three volcanic cones: Kibo, with a height of 5895 m the highest peak in Africa, Mawenzi and Shira. When the ISS flew over Kilimanjaro, the dense cloud cover had just revealed a view of its snow-covered peak. ISS, July 23, 2018. (Photo: ESA/NASA, Alexander Gerst)

87 The Andes, with a length of about 9000 km, form the longest single mountain range on Earth. Their formation is due to the subduction of the Nazca lithospheric plate of oceanic crust under the South American Plate of continental crust, known as "subduction". Through this process, material in both plates melts and makes its way to the surface in a volcanic eruption. Along the west coast of South America, therefore, one finds numerous volcanoes, which can be seen here in the photographs of ESA astronauts Alexander Gerst (above) and Samatha Cristoforetti (below). They form part of the Pacific Ring of active volcanoes, also known as the "Pacific Ring of Fire". ISS, June 30, 2018 and August 28, 2022. (Photos: ESA/NASA, Alexander Gerst and Samantha Cristoforetti)

88 The Kuril Islands chain in Russia also hosts numerous active volcanoes. The cause of this is again the tectonic behavior of the lithosphere. The northwestward pushing Pacific Plate subducts here under the Okhotsk Plate. Just like in the case of the Andes, volcanoes are formed on the surface by the melting rock. Here, the Sarychev volcano on the Matua Island is erupting in June 2009. ISS astronauts have captured this eruption. An eruption cloud of brown ash and white steam reached up to heights of 14 km. Above it, condensation in the rapidly rising air has created a pileus cloud, while on the southwest flank of the volcano a pyroclastic moves downhill. ISS, June 12, 2009. (Photo: Earth Science and Remote Sensing Unit, NASA Johnson Space Center, ISS020-E-9048)

89 In September 1994, the Rabaul in Papua New Guinea erupted. Its ash cloud rose to a height of almost 20 km. In the image of the 64th shuttle mission, the ash can be recognized by the brownish color of the cloud top. A second, also brownish ash cloud remained at low altitudes due to low winds and spread there. The eruption of Rabaul destroyed parts of the namesake city, which had been evacuated in time. The cloud-covered island in the foreground is New Ireland, located east of Rabaul. STS64, September 19, 1994

(Photo: Earth Science and Remote Sensing Unit, NASA Johnson Space Center, STS064-116-64)

90 Florida, the Bahamas, and the Gulf of Mexico have been a popular subject of images since the beginning of manned space travel, especially because of the interplay of differently deep waters and land in the form of numerous islands. These motifs attracted every crew, including that of the 92nd shuttle mission in 1998. On board was the then 77-year-old John Glenn, who had already viewed the region 36 years earlier from the perspective of the Mercury space capsule as America's first astronaut. STS95, October 31, 1998. (Photo: Earth Science and Remote Sensing Unit, NASA Johnson Space Center, STS095-743-033)

91 The Great Barrier Reef off the coast of Australia is considered a unique ecosystem. Corals have created a 2300 km long structure of individual reefs and islands, of which almost 200 km can be seen in this image. ISS, November 16, 2018. (Photo: ESA/NASA Alexander Gerst)

92 Lake Nasser in Egypt shines in the light of the high-standing Sun. The damming of the Nile in Aswan has created a 500 km long reservoir. Its waterline is so high that the lake extends into the side valleys of the mountains to the east. ISS, January 23, 2005. (Photo: Earth Science and Remote Sensing Unit, NASA Johnson Space Center, ISS010-E-14618)

93 The Niagara Falls on the border between the USA and Canada are among the largest waterfalls in the world. The Niagara River, coming from Lake Erie, transports 2.8 million liters per second over the waterfalls into Lake Ontario. Goat Island divides the river into two arms. To the left of this island, the larger Horseshoe Falls can be seen, to the right the American Falls and Bridal Veil Falls. ISS, April 2, 2021. (Photo: Earth Science and Remote Sensing Unit, NASA Johnson Space Center, ISS064-E-53177)

94 Manned space flights cannot provide us with images of the polar caps, as their orbits, unlike those of automatic platforms, do not orbit perpendicular to the equator. Space Shuttle flights reached a maximum northern and southern latitude of 57°. From there, however, at least a view of the southern tip of Greenland with Cape Farvel was possible. At the end of March, it appeared cloudless and snow-covered. On the eastern side, the East Greenland Current forms patterns from ice and drives it south. STS45, March 29, 1992. (Photo: Earth Science and Remote Sensing Unit, NASA Johnson Space Center, STS045-152-105)

95 In Patagonia, we find the largest contiguous glacier area in the southern hemisphere. Air currents approaching from the south bring rain and snow and maintain a now 13000 km2 large ice field. ISS, February 13, 2014. (Photo: Earth Science and Remote Sensing Unit, NASA Johnson Space Center, ISS038-E-47324)

96 Sunrise over the Philippine Sea, northeast of Manila. The cloud towers in the tropical atmosphere cast long shadows. This image gives an impression of the Earth's atmosphere, which is constantly in motion. ISS, August 19, 2020. (Photo: Earth Science and Remote Sensing Unit, NASA Johnson Space Center, ISS063-E-76166)

97 Low-pressure vortices are characterized by a dense, counterclockwise rotating cloud spiral. Hurricanes are extreme cases. Hurricane Isabel was already seen in the previous chapter in perfect top view from 700 km height. In contrast, the ISS was only 380 km high when it took a slanted look into this gigantic hurrican. The image provides a close-up of its eye. ISS, September 15, 2003. (Photo: Earth Science and Remote Sensing Unit, NASA Johnson Space Center, ISS007-E-14886)

98 The highest clouds found in the Earth's atmosphere are the so-called "noctilucent clouds" at a height of 85 km above the Earth's surface. They consist of ice crystals and primarily form around the time of the summer solstice in high northern and southern latitudes. The photographer on the ISS captured this type of cloud after the Sun disappeared behind the horizon like thin cirrus. ISS, July 22, 2008. (Photo: Earth Science and Remote Sensing Unit, NASA Johnson Space Center, ISS017-E-11632)

99 Dust storms are a rewarding photo subject for astronauts. In this photo, as the International Space Station was just over Libya, Alexander Gerst's view to the south over the Sahara was obscured by an extensive dust storm. White cumulus and cumulonimbus clouds tower over the yellowish-brown colored dust storm. ISS, August 8, 2014. (Photo: Earth Science and Remote Sensing Unit, NASA Johnson Space Center, ISS040-E-90343)

100 Both night shots were taken just a few minutes apart. They show two atmospheric phenomena. In each image with the Earth's horizon against the dark night sky, the faint greenish airglow appears, which continuously surrounds the Earth at a height of 100 km. It is mainly caused by the ultraviolet radiation of the Sun, which ionizes atoms and molecules of the upper atmosphere during the day. During the transition to the original state, the so-called "recombination" weak light is emitted, which can also be seen in the night sky after sunset. In addition, the so-called "red sprites" can be seen as red parallel structures above the glow of a thunderstorm, a very rarely observed thunderstorm phenomenon. These luminous signatures are cold discharges that strike upwards in the atmosphere above a thunderstorm up to heights of 100 km. At the top, the red sprites appear more than 2000 km away over Missouri or Illinois to the left of the bright Moon, which is located just above the night sky glow. In the lower picture, you can see thunderstorm lights on the right and above them the red structures of the "red sprites" over Honduras. ISS, August 10, 2015. (Photo: Earth Science and Remote Sensing Unit, NASA Johnson Space Center, ISS044-E-45553/45576)

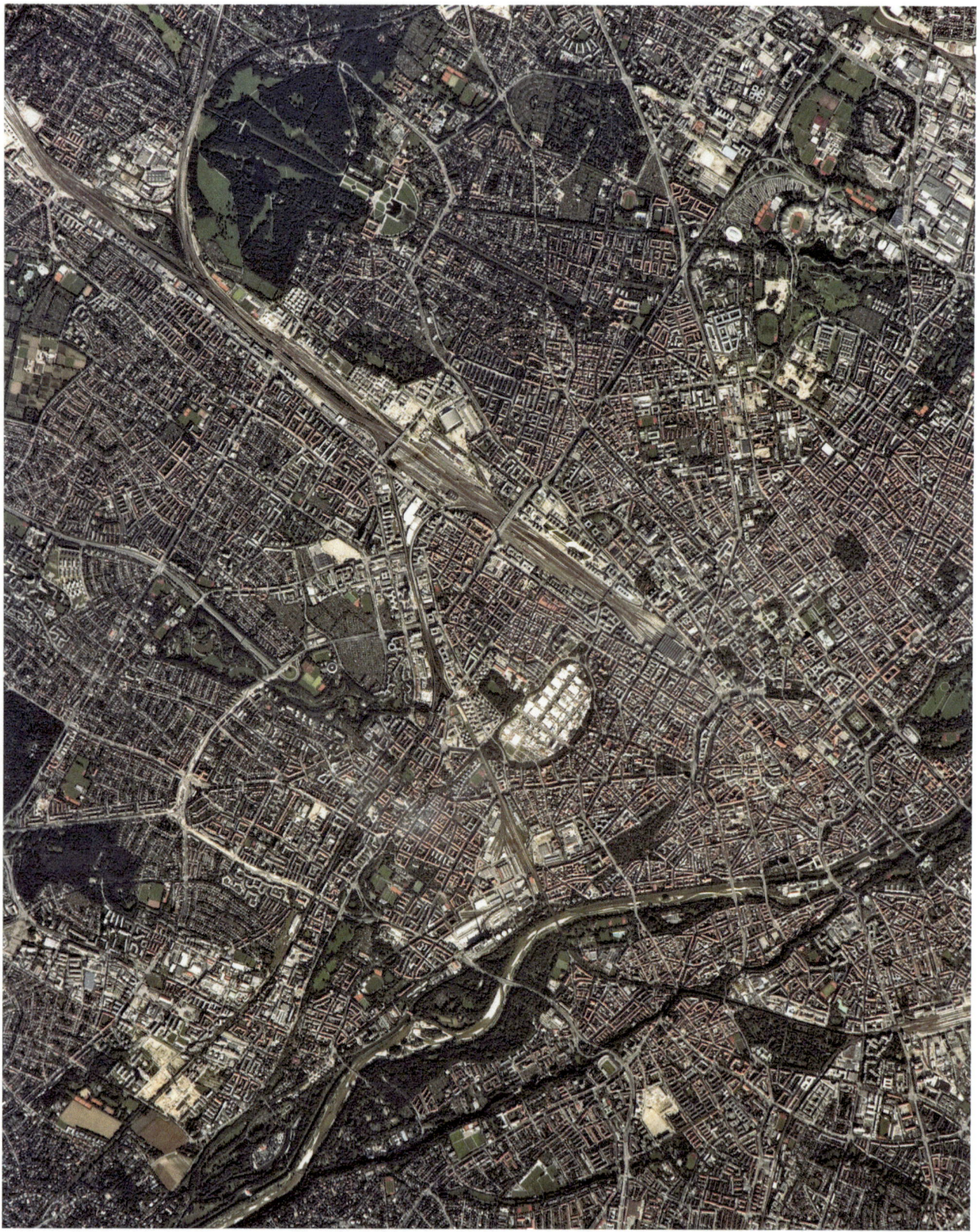

101 Large cities appear from space as distinctive patterns of our civilization. Munich shown here appears, due to the extreme 1200 mm telephoto lens used on the camera, as if taken from a commercial aircraft even from a distance of 382 km. In the center of the image, you can see the oval of the Theresienwiese with the Oktoberfest taking place. ISS, September 24, 2011. (Photo: Earth Science and Remote Sensing Unit, NASA Johnson Space Center, ISS029-E-9235)

102 Two views from the ISS of New York, as the space station was about 400 km above the Earth. In both shots, you can see the rectangle of Central Park. On the left, the shadows of the skyscrapers even make the image appear three-dimensional. A year later, a night view of the same scene was achieved (right). ISS, May 5, 2014 and October 18, 2015. (Photo: Earth Science and Remote Sensing Unit, NASA Johnson Space Center, ISS039-E-18512/ISS045-E-66112)

103 The International Space Station's orbit also takes it past the east coast of the United States. In February 2012, this happened at the end of a night and the lights of the major cities New York with Long Island (right), Philadelphia, Baltimore and Washington as well as Richmond and Norfolk (left) lined up along a more than 800 km wide arc. ISS, February 6, 2012 (Photo: Earth Science and Remote Sensing Unit, NASA Johnson Space Center, ISS030-E-78095)

104 The Iberian Peninsula during a flyover in 2014. Lisbon and Porto stand out on Portugal's coast. In Spain, the centrally located Madrid and, to the south near the coast, Seville and Granada are prominent. Above the horizon, the greenish airglow glow of the night sky curves. ISS, July 26, 2014 (Photo: Earth Science and Remote Sensing Unit, NASA Johnson Space Center, ISS040-E-81320)

105 A night shot from the International Space Station of the Korean Peninsula. As the ISS flew over the East China Sea in January 2014, an astronaut looked north. In the lower right half of the image, South Korea extends with the brightly lit Seoul, above it along the 38th parallel is the border to North Korea. What appears as a dark sea connecting the Sea of Japan in the east and the Yellow Sea in the west is the darkness-enveloped North Korea with its capital Pyongyang as a brighter spot of light. ISS, January 30, 2014. (Photo: Earth Science and Remote Sensing Unit, NASA Johnson Space Center, ISS038-E-38300)

106 The full Moon illuminates this scene of southern Scandinavia from the International Space Station. The light of our Moon illuminates snowfields in southern Norway and cloud fields over the Baltic Sea and Finland. Large cities like Copenhagen, Malmö, Gothenburg or Stockholm are easy to identify. On the horizon, the green glow of the Aurora Borealis is visible. ISS, April 3, 2015. (Photo: Earth Science and Remote Sensing Unit, NASA Johnson Space Center, ISS043-E-86375)

107 Also illuminated by the full Moon is the eastern Mediterranean area with Israel, Egypt's Nile Delta, and further north Cyprus and on the horizon Turkey. Tel Aviv on Israel's coast, east of it Amman in Jordan and Cairo at the southern end of the Nile Delta are the brightly lit metropolises in this scene. ISS, September 16, 2016. (Photo: Earth Science and Remote Sensing Unit, NASA Johnson Space Center, ISS049-E-4516)

108 Here the ISS looks at the John F. Kennedy Space Center on Merritt Island. In the front directly on Florida's east coast, you can see the Cape Canaveral Space Force Station and in between the lagoon of the Banana River. ISS, December 27, 2020. (Photo: Earth Science and Remote Sensing Unit, NASA Johnson Space Center, ISS064-E-015920)

109 What Cape Canaveral is for the USA, is for Russian space travel the Baikonur Cosmodrome located in Kazakhstan. Since 1957, many of the Soviet and Russian space missions have launched from there. This is also where many of the US and ESA astronauts took off for their stays on the International Space Station. ISS, October 11, 2018. (Photo: Earth Science and Remote Sensing Unit, NASA Johnson Space Center, ISS057-E-23187)

110 For an observer from near-Earth orbit, impact craters are not always easy to recognize. Even if no clouds obstruct the direct view, the often heavily eroded remnants of the impact are hardly identifiable. Only when they appear as a simple, small impact crater, like here the Roter Kamm crater in Namibia's Namib Desert with a diameter of 2.5 km, do they stand out. Its age is estimated to be less than 5 million years. ISS, October 14, 2006. (Photo: Earth Science and Remote Sensing Unit, NASA Johnson Space Center, ISS014-E-5597)

111 From the impact structure Gosses Bluff in Australia with a diameter of 22 km, at first glance, only the remnant of the always occurring central elevation, the "central uplift", as a ring-shaped hill chain is visible. Gosses Bluff is dated to an age of just over 140 million years. ISS, February 6, 2007. (Photo: Earth Science and Remote Sensing Unit, NASA Johnson Space Center, ISS014-E-13613)

112 Since the 1960s, Manicouagan in Quebec, Canada, a 100 km large impact, has been marked by a dammed ring- shaped water reservoir. In February 2003, the International Space Station flew over northern Canada and captured this winter scene with the frozen Manicouagan crater west of the St. Lawrence River. ISS, February 26, 2003. (Photo: Earth Science and Remote Sensing Unit, NASA Johnson Space Center, ISS006-E-34153)

113 On their journey around the Earth, astronauts pass the 90 km large Acraman impact structure in Australia, marked by large salt lakes and evaporation basins like the small round Lake Acraman visible near the center of the image and the bright structure of Lake Gairdner to the right. Like Manicouagan, Acraman is one of the few impact craters visible from low Earth orbit. ISS, October 2, 2014. (Photo: Earth Science and Remote Sensing Unit, NASA Johnson Space Center, ISS041-E-61822)

114 Polar lights also fascinate astronauts. They can direct their cameras towards the horizon where these interesting phenomena occur. Reddish glow comes from oxygen atoms at 200 km altitude when electrons from the solar wind are captured by the Earth's magnetic field and ionize the atoms. A green aurora is created lower at about 100 km altitude, also by the ionization of oxygen. This division is exactly what the "Northern Lights", the Aurora Borealis over North America, illustrates. On the ground, the illuminated cities and towns of Montana and Alberta can be seen. ISS, February 19, 2012. (Photo: Earth Science and Remote Sensing Unit, NASA Johnson Space Center, ISS030-E-110559)

115 The "Southern Lights" the Aurora Australis, over the southern Indian Ocean. On the horizon is the constellation Orion, recognizable by the belt and sword stars. ISS, September 17, 2011. (Photo: Earth Science and Remote Sensing Unit, NASA Johnson Space Center, ISS029-E-6020)

116 Our home planet is constantly colliding with small objects in the solar system. Most of them are only the size of a dust grain. During the collision, the particles are slowed down by the Earth's atmosphere and cause air molecules to glow. At certain times, the Earth passes through regions with higher dust density on its orbit. Then meteors occur more frequently. One of the most well-known meteor showers are the Perseids in early August, which are due to the slow dissolution of comet 109P/SwiftTuttle. Such a Perseid caught the attention of an ISS astronaut in August 2011, who captured its short luminous trail as it entered the Earth's atmosphere. ISS, August 13, 2011. (Photo: Earth Science and Remote Sensing Unit, NASA Johnson Space Center, ISS028-E-024847)

117 A solar eclipse is an interesting phenomenon, even when observed from Earth orbit. When the Moon's umbra moves across the Earth, there is a strange darkness during the day. In August 2017, ESA astronaut Paolo Nespoli followed the eclipse spectacle over the USA from the ISS at a height of 400 km. On the horizon, you can see the dark region of the umbra. ISS, August 21, 2017. (Photo: ESA/NASA

118 In low Earth orbit, an observer of the Earth's Moon is practically above the entire atmosphere and can unobstructedly follow how the Moon rises from our air envelope. One can see how the decreasing density of the air layers towards the top makes the Moon appear strongly deformed due to the density-dependent light refraction when it first appears. The light rays at the lower edge of the Moon experience a stronger refraction than those at the upper edge. As the Moon rises higher, this differential light refraction has less and less effect and our satellite regains its round shape. ISS, August 18, 2016. (Photo: Earth Science and Remote Sensing Unit, NASA JohnsonSpace Center, Jeff Williams, ISS048-E-61100 to 61111)

119 Always beautiful to see is the Moon standing above the Earth's horizon. ISS, September 4, 2010. (Photo: EarthScience and Remote Sensing Unit, NASA Johnson Space Center, ISS024-E-13421)

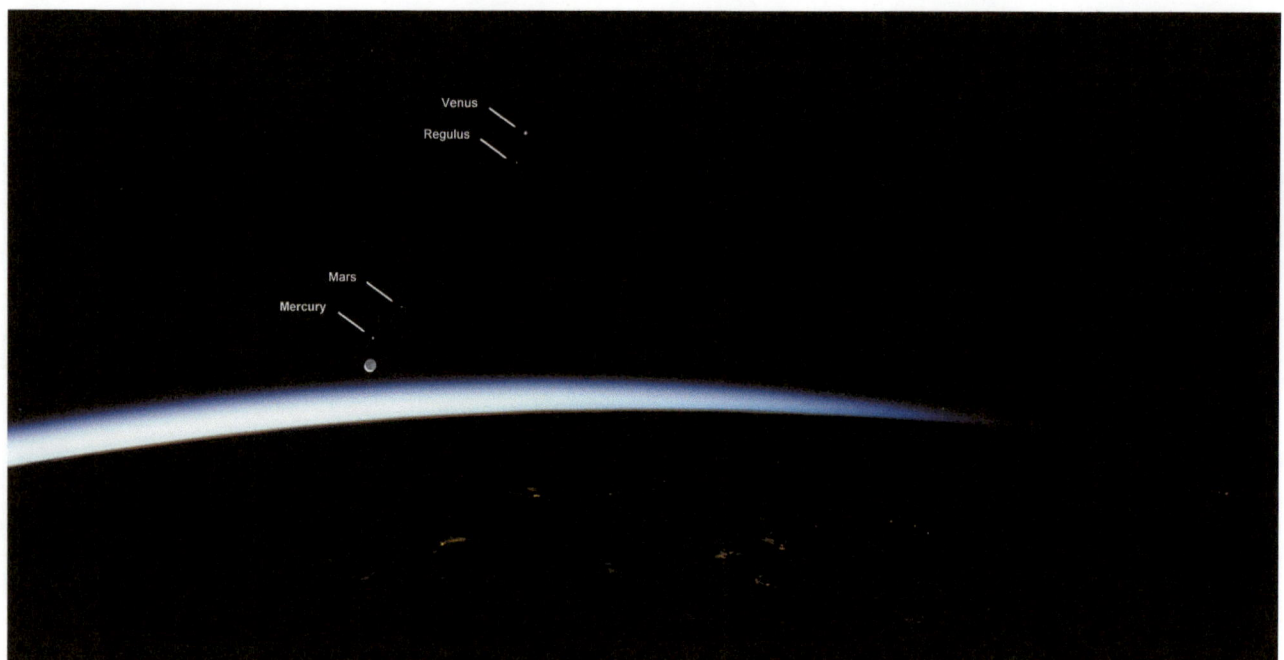

120 Here you can see from bottom to top a series of celestial objects above the Earth's atmosphere: Moon, Mercury, Mars, Regulus (brightest star in the constellation Leo) and Venus. ESA astronaut Paolo Nespoli took this picture from the International Space Station. ISS, September 18, 2017. (Photo: ESA/NASA)

121 Bright comets are rare. As long as manned space travel had to do without permanent stations in space, it was hardly possible to track a comet from near-Earth space during a short-term space flight. In the mid-90s, Hale-Bopp, a long-period comet, appeared, which was not only very bright, but also stayed in the sky for a long time. During its phase of greatest brightness, the 83rd shuttle mission took place in early April 1997. Hale-Bopp was impossible to overlook from the spacecraft. The crew took the first picture of a comet standing above the atmosphere and the night glow. STS83, April 1997. (Photo: Earth Science and Remote Sensing Unit, NASA Johnson Space Center, STS083-507-023)

122 From the ISS, it was possible to observe the briefly visible comet Lovejoy C/2011 W3 in December 2011. Its long tail was particularly impressive. ISS, December 22, 2011. (Photo: Earth Science and Remote Sensing Unit, NASA Johnson Space Center, ISS030-E-15472)

123 When stars rise or set on the horizon due to the movement of the ISS, they are only dimmed by the airglow. Here, the constellation Southern Cross stands directly above the edge of the Earth. ISS, February 28, 2003. (Photo: Earth Science and Remote Sensing Unit, NASA Johnson Space Center, ISS006-E-51890)

124 A surprise guest in the summer of 2020 was the comet Neowise. It developed into a comet visible to the naked eye. In this picture, you can see the comet just above the Earth's atmosphere, which already reveals the daylight side of the orbit during the flight of the International Space Station towards the east. ISS, July 5, 2020. (Photo: Earth Science and Remote Sensing Unit, NASA Johnson Space Center, ISS063-E-39827)

125 With modern high-sensitivity digital cameras, even star field images from Earth orbit are possible, which resemble astro images from the ground. The southern Milky Way extends in this photo over 100° from the constellations Vela (left) to Norma (right) with the dark cloud of the Coal Sack to the right of the center of the image. ISS, July 23, 2020. (Photo: Earth Science and Remote Sensing Unit, NASA Johnson Space Center, ISS063-E-54340)

126 A view from the ISS to the center of our galaxy, the Milky Way, obscured by dark clouds. On the left, it connects to the Milky Way in the constellation Norma of the previous image. Different colored layers of the night sky glow can be seen. The bright region in the atmosphere shows the location of a thunderstorm over the Pacific. ISS, August 9, 2015. (Photo: Earth Science and Remote Sensing Unit, NASA Johnson Space Center, ISS044-E-45215)

6. Geostationary Orbit - A Whole View of Earth

If you want to see the entire Earth as a whole, floating freely, you have to move your location from the low altitudes of low Earth orbits further into space. From many 1000 km, the Earth begins to fill a significantly smaller part of the field of view. Two types of orbits have established themselves for satellites at such great heights – the geostationary orbit and elongated elliptical orbits. In the latter, a satellite orbits the Earth on a highly elliptical path. The point closest to Earth (perigee) is a few hundred kilometers above the Earth's surface, the farthest (apogee) several 10,000 kilometers. According to Kepler's laws, a satellite on this path moves fastest near the Earth and slowest at the apogee. Therefore, in the Earth-distant part of the orbit, the Earth remains in view for a long time.

From such an orbit, NASA tried to obtain the first photo of the Earth with Explorer-6 launched on August 7, 1959, in the very early beginnings of space travel. A simple TV system on Explorer-6 was supposed to provide the first images of cloud cover. To do this, a strip-wise scanning of the Earth's surface was attempted using the forward movement of the satellite in orbit and its spin motion, which stabilized its position in the orbit. On August 14, at a distance of 27,000 km, the light sensor swept over the Earth and delivered an image of the cloud cover over the Pacific Ocean. The scientific final report of NASA on Explorer-6 roughly states: "While the TV system is described in the scientific literature, the quality of the results obtained with it was not sufficient to present them there." Nowadays, images of this quality immediately end up in the computer's trash bin, but the first failed Earth shot from Explorer-6 can still be found in NASA's archives.

Elongated elliptical orbits were particularly interesting for the former Soviet Union. Due to the high northern latitudes of a large part of their territory, satellites standing over the equator were only suitable to a limited extent to ensure communication for the entire area. A highly elliptical orbit, on the other hand, was different. If one chose it so that its apogee was above the Northern Hemisphere, a satellite remained visible over the horizon for hours. This concept was successfully implemented for the first time with the launch of Molnija-1 in 1965. In 12 hours, the satellite orbited the Earth, moved up to 40,000 km in the apogee, and remained visible over the Northern Hemisphere for about 8 hours. The third Molnija launch also carried a black-and-white camera, which on May 30, 1966, transmitted the first known, relatively coarsely resolved image of an Earth illuminated about half. However, it was hardly noticed by the public at the time. Russia continues to use such highly elliptical orbits to establish communication possibilities for its far northern territory and to probe it meteorologically. The recently launched first satellite of the Arktika series provides high-resolution images from a completely new perspective, standing far above the North Pole. The photos obtained with modern sensors show a level of detail comparable to the images from geostationary satellites.

Little attention was also paid to the color photos from an unmanned test flight of the Saturn V rocket together with the Apollo command and service module more than a year later. Under the designation Apollo 4, the Saturn V was sent on an elliptical orbit with a maximum Earth distance of 18,000 km in November 1967. A film camera was located in the command module, which photographed the Earth in color from distances of 13,500-18,000 km. After the command unit splashed down in the Pacific, the films could be recovered and hundreds of photos evaluated. Unlike the images transmitted by automatic TV cameras until then, they showed the Earth in a natural blue-white aesthetic.

The view of a "full Earth" from a highly elliptical orbit remained the exception. Far more often, images of the Earth are found from geostationary satellites. They have evolved from their beginnings in the 1960s into indispensable helpers in the fields of communication and meteorology. How long a probe orbiting the Earth takes for one orbit depends on its orbital height. At the minimum possible height of 200 km, an orbit only takes 88 minutes, at 1000 km already 105 minutes, and at 35786 km above the Earth's surface with 23 hours 56 minutes 4 seconds exactly the length of a day, i.e., one Earth rotation. If the satellite moves at this height with a certain inclination to the Earth's equator we speak of geosynchronous orbits. Relative to the Earth, it covers an area north and south of the equator, which has the shape of a so-called "analemma". When the inclination reaches the value 0°, the satellite appears to stand still to an observer on the ground; it orbits the Earth in a geostationary orbit. All satellites that have such orbits are located quite exactly above the equator. Depending on which longitude they were placed, they can continuously observe a certain hemisphere. From such an orbit, the Earth appears at an angle of about 17°.

After NASA had already successfully placed a telecommunications satellite in a geostationary orbit in 1964, ATS-1, abbreviated for "Application Technology Satellite No. 1" followed in December 1966. In addition to numerous technological test requirements for operating a satellite at this height, ATS-1 was also to photograph the global cloud cover for the first time. For this purpose, a special black-and-white camera was on board. ATS-1 was placed in a geosynchronous orbit with an initial position over the eastern Pacific. From there, shortly after the start, the first high-quality images of the "whole Earth" were made. Although it was no longer a real premiere due to the Molnija recording that had been made months before, it was the first time that the weather events over the eastern Pacific area and the American continent could be followed in detail. A short time later, in July 1967, the DODGE satellite developed by the American Department of Defense followed in a not exactly geosynchronous orbit - the orbital height was only about 33500 km; thus, an orbit around the Earth only took 22 hours. Like ATS-1, its tasks were primarily to test then new technologies. It also carried two cameras; one of them even allowed color photographs. Although they were mainly used to provide information about the orientation of the satellite, the Earth could also be captured. In this way, the first color photo of the complete Earth was taken in August 1967. Little notice was taken of it; because shortly afterwards, in November of the same year, NASA brought ATS-3, a further development, into the geosynchronous orbit with

a low orbital inclination. Positioned over the West Atlantic, it delivered a color image with much better resolution in the same month.

Once it was shown that the dynamics of the weather events can be followed from about 36000 km height, the construction of a system of geostationary weather satellites began in 1974. NASA first sent the first two such satellites into space and placed them over the equator with a view of the west Atlantic and east Pacific area. They were later replaced by a series of GOES satellites, a synonym for "Geostationary Operational Environmental Satellites". Today they are operated by NOAA, the US organization for monitoring the oceans and the atmosphere. The then USSR also sent its first geostationary satellite on its 24-hour orbit in 1974. The European Space Agency ESA positioned its first satellite Meteosat exactly on the prime meridian over Africa in 1977. It was the beginning of a series of more than 10 further Meteosat platforms. In Europe, too, EUMETSAT, an organization specifically founded for satellite meteorology, is now responsible for these satellites. responsible. Also in 1977, Japan began to constantly monitor the meteorology of the Western Pacific hemisphere. In the following years, a system of geostationary meteorological platforms developed, in which practically all space-faring nations - China, India, and Korea - participated. It covers the entire geographical longitude range and guarantees complete monitoring of global weather events. Those satellites that are not currently needed are in standby mode and can be maneuvered to another position in the geostationary orbit if needed, for example, after the failure of another satellite.

The main payload of the geostationary meteorological satellites consists of sensors that register the radiation coming from the Earth in the visible and infrared spectral range. This results in images of the Earth from a height of 36000 km every 15-30 minutes. If you compare the images of the individual satellite operators, you will find striking differences. Some combine the spectral information with the aim of obtaining as realistic-looking images as possible; others try to highlight certain details, such as clouds. The spatial accuracy of the images reaches up to 1 km. From the rapidly successive series of images, one can follow the constantly changing weather events. The information obtained is indispensable for reliable weather forecasting today and is indespensible to follow the changing weather on Earth.

But the fixed position of the satellite relative to the Earth also allows vivid insights into celestial mechanical processes that are related to the daily rotation of the Earth and its annual orbit around the Sun. We see the daily course of the Sun; how it first illuminates the eastern part of the Earth's disk and then, with the terminator moving further and further west, finally immerses the entire hemisphere in Sunlight at noon. Afterwards, night falls on the eastern edge of the Earth and spreads westward until after twelve hours the entire Earth is plunged into darkness and the sequence begins anew. Seasonal changes are also reflected in the lighting conditions on Earth. In the morning and evening, the satellite, Earth, and Sun are at right angles to each other. Then the position of the Earth's axis is shown at the terminator. At the time of the spring and autumn equinox, it is vertical, at the times of the summer or winter solstice, it is inclined with stronger illumination of the Arctic or the Antarctic. During the solstice, one can even identify the midnight Sun or the polar night in the images of the geostationary satellites. The latter appears in recordings at noon, when regions with high northern or southern latitude are no longer illuminated. The midnight Sun, on the other hand, appears as a narrow bright crescent around the North or South Pole.

The permanent visibility also allows the tracking of phenomena that can only be randomly captured by satellites in low orbits - they need several days to cover the entire globe. These include solar eclipses. From geostationary satellites, however, the shadow of the Moon can be tracked over long distances of its path on the Earth's surface.

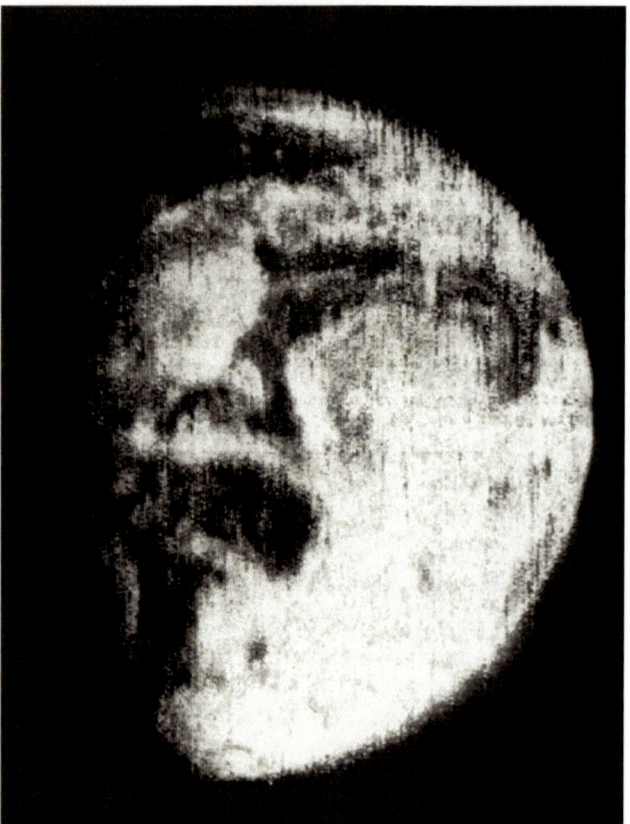

127 A historical photo, but without image content: The first publicly known attempt to image the Earth from space. In 1959, it failed for Explorer-6. The image only shows a diffuse bright structure that was created when scanning the Earth from a height of 27000 km. Explorer-6, August 14, 1959. (Photo: NASA)

128 The first known image of a half-illuminated Earth, taken from more than 10000 km away by a Russian Molnija satellite in May 1966, which stood on its elliptical orbit over the North Pole. Molnija-1, May 31, 1966. (Photo: NASA)

129 The color image of the partially illuminated Earth, as captured more than a year later by Apollo 4 during the first unmanned test flight of the Saturn V rocket by a 70 mm camera housed in the Apollo command module, has a completely gives a completely different color impression. The image from a distance of 15,000 km resembles the later images taken during Moon flights. Apollo 4, November 9, 1967. (Photo: NASA)

130 Two images of the Earth, also from the mid-60s. The primary purpose of ATS-1 was to test TV transmission from geosynchronous orbits. In addition, it also provided images in which a large part of an Earth hemisphere was visible, as here in one of the first images from December 1966 (left). They demonstrated the potential of satellites in such orbits for meteorological observations. ATS-1 also produced the first image of the Earth-Moon system, which presented both celestial bodies more than 50% illuminated (right). Even here, the strong brightness difference between Earth and Moon was noticeable. ATS-1, 11th and 22nd December 1966 (Photo: NASA, University of Wisconsin-Madison)

131 The first color image of the full Earth was a byproduct of the DODGE mission, a joint project of NASA and the US Air Force. Methods for aligning satellites in geosynchronous orbits were investigated and the Earth was also photographed. Coarse color images were obtained using filters. DODGE, August 1967. (Photo: NASA, US Air Force, Johns Hopkins University Applied Physics Laboratory)

132 The color images that the now already "experimental geostationary weather satellite" named ATS-3 transmitted to Earth shortly after the DODGE mission were significantly better. With higher resolution and natural color rendering, it provided a realistic image of our planet. ATS-3, November 10, 1967. (Photo: NASA, University of Wisconsin-Madison)

133 In the mid-70s, the construction of a global system for permanent monitoring of global weather began. GOES-1, the first continuously operating geostationary satellite of this kind, started its activity in October 1975. Standing over the equator, it had both American continents in view and regularly sent images of the western hemisphere from there. GOES-1, October 25, 1975. (Photo: NOAA)

134 From December 1977, Europe also had the capabilities to monitor weather events by means of a geostationary satellite, positioned over Africa. Meteosat-1, December 9, 1977. (Photo: European Space Agency - ESA)

135 Meteorological satellites register the radiation coming from the direction of the Earth, i.e., scattered sunlight and thermal emission in different wavelength ranges. From these data, meaningful image products are created for meteorological services and science, which are not necessarily beautiful images of our home planet in natural colors. To create the latter requires some data technical acrobatics. Each meteorological weather satellite facility has its own procedures for this, which are adapted to the peculiarities of the respective instruments. Therefore, the colored Earth images from different geostationary satellites only partially resemble each other.

For Europe, the European organization EUMETSAT always keeps a satellite ready at 0° geographical longitude over Africa. Above you can see the Earth in an image of the third satellite of the second Meteosat generation, taken about an hour before noon. The image was created from the data of different spectral channels in a possible natural color rendering. The subtropical latitudes appear largely cloud-free while dense cloud cover can be seen over the equator. An Atlantic low is approaching the European continent.

The lower photo comes from the youngest offspring of the Meteosat family, Meteosat-12, which is now the third generation of Meteosat satellites. Meteosat-9 and Meteosat-12, August 7, 2012 and March 18, 2023. (Photos: EUMETSAT/ESA)

136 The view of the US eastern GOES satellite, which was placed at 75° west longitude, shows the two American continents. Here too, when combining the spectral channels, a design was chosen that corresponds as closely as possible to the visual impression. Hurricane Earl is approaching the east coast of the United States. GOES-13, September 3, 2010. (Photo: NASA, GSFC, NOAA)

137 GOES-16 is currently performing the task of the eastern American geostationary satellite at a geographic longitude of 75°. Its imaging instrument is more powerful than that of its predecessor GOES-13. This can be seen in this full Earth image from 2017. GOES-16, January 15, 2017. (Photo: NOAA)

138 The image of the Russian satellite Elektro-L at 76° east longitude appears differently again. Elektro-L, September 22, 2013. (Photo: Roscosmos, Vitaliy Egorov, CC BY 3.0)

139 Far to the east over the Pacific at 140° east longitude is Himawari-8, a Japanese weather satellite. Its field of view covers the Japanese islands, Australia, and the western Pacific region. Here too, close to the visual impression, a slightly different color scheme is shown. The image from Himawari-8 resembles Earth images as captured by Apollo astronauts on Moon flights with medium format cameras. Himawari-8, February 10, 2016. (Photo: NOAA, Data: JMA – Japanese Meteorological Agency)

140 The image of the Earth in a current image of the Russian satellite Arktika-M1 looks completely different from the photo taken 55 years earlier in Figure 128. From its highly elliptical orbit, it took this picture a few weeks after its launch at the end of February 2021 from a completely unfamiliar perspective high above the Arctic. Greenland is above the center of the image. The sensor of Arktika M1 largely corresponds to that of Elektro-L. The photo shows a level of detail that we otherwise know from geostationary satellites. Arktika-M1, March 26, 2021 (Photo: Roscosmos, Russian Space Systems, NPO Lavochkin, Planeta Research Center for Space Hydrometeorology, Roshydromet)

141 Weather satellites in geostationary orbit not only provide images of the entire Earth. Their sensors also allow insight into regional weather events. In Europe, the summer of 2003 was considered the summer of the century. The high pressure system Michaela brought particularly Central and Southern Europe extended periods of cloudless skies as here in August at the height of the heatwave. Meteosat-8. August 10, 2003. (Photo: EUMETSAT)

142 An impressive panorama of the Himalayas with the Tibetan Plateau to the north and the Ganges Plain to the south, taken by the Russian weather satellite Elektro-L. Elektro-L, July 23, 2013. (Photo: Roscosmos, Vitaliy Egorov, CC BY 3.0)

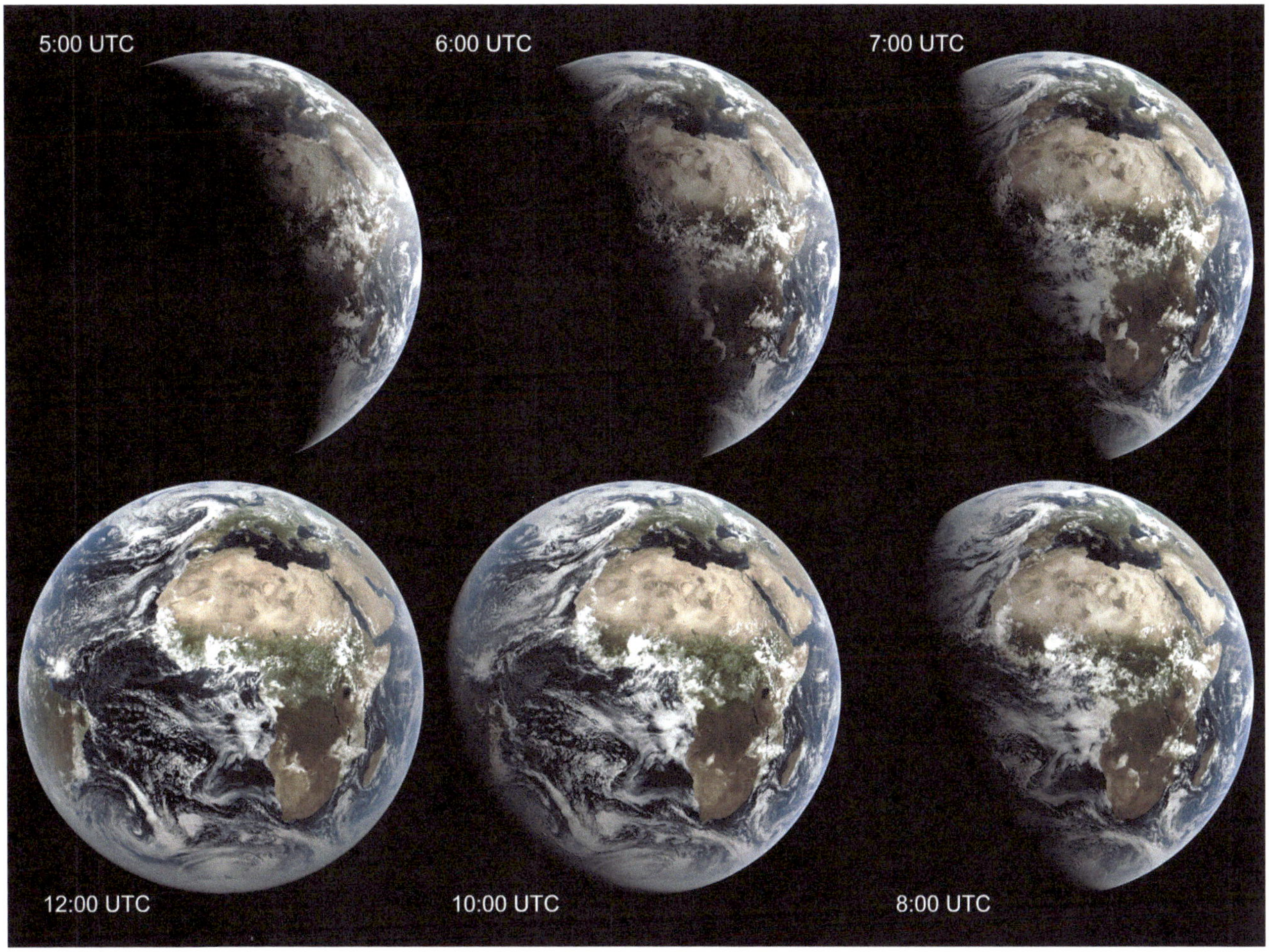

5:00 UTC   6:00 UTC   7:00 UTC

12:00 UTC   10:00 UTC   8:00 UTC

143 Seven hours from the perspective of the EUMETSAT satellite Meteosat-9, which is located above the equator. The image starts in the top left at 5:00 UTC with a partially illuminated Earth crescent. At noon (bottom left), the Earth is then fully illuminated. This image was created using a special process, which enabled its special, very natural-looking impression. Meteosat-9, July 21, 2009. (Photo: Maximilian Reuter, IUP-IFE, University of Bremen, Data: EUMETSAT)

March 21, 2011   June 21, 2011   September 23, 2011   December 21, 2011

144 At 6:00 UTC every day, the Sun, Earth, and geostationary satellite form a right-angled triangle and the Earth appears half illuminated. The terminator illustrates how the Earth's axis is tilted relative to the Sun and it allows us to track the seasonally changing lighting conditions on Earth. At the equinoxes, in March (left) and September (third image), the Earth's axis does not point towards or away from the Sun. The northern and southern hemispheres are equally illuminated. At the summer solstice in June (second image), the northern hemisphere is tilted towards the Sun and receives more light. At the winter solstice in December (right image), the situation is exactly reversed. Then the southern hemisphere points towards the Sun. Meteosat-9, 2011. (Photo: Maximilian Reuter, IUP-IFE, University of Bremen, Data: EUMETSAT)

110

6. Geostationary Orbit - A Whole View of Earth

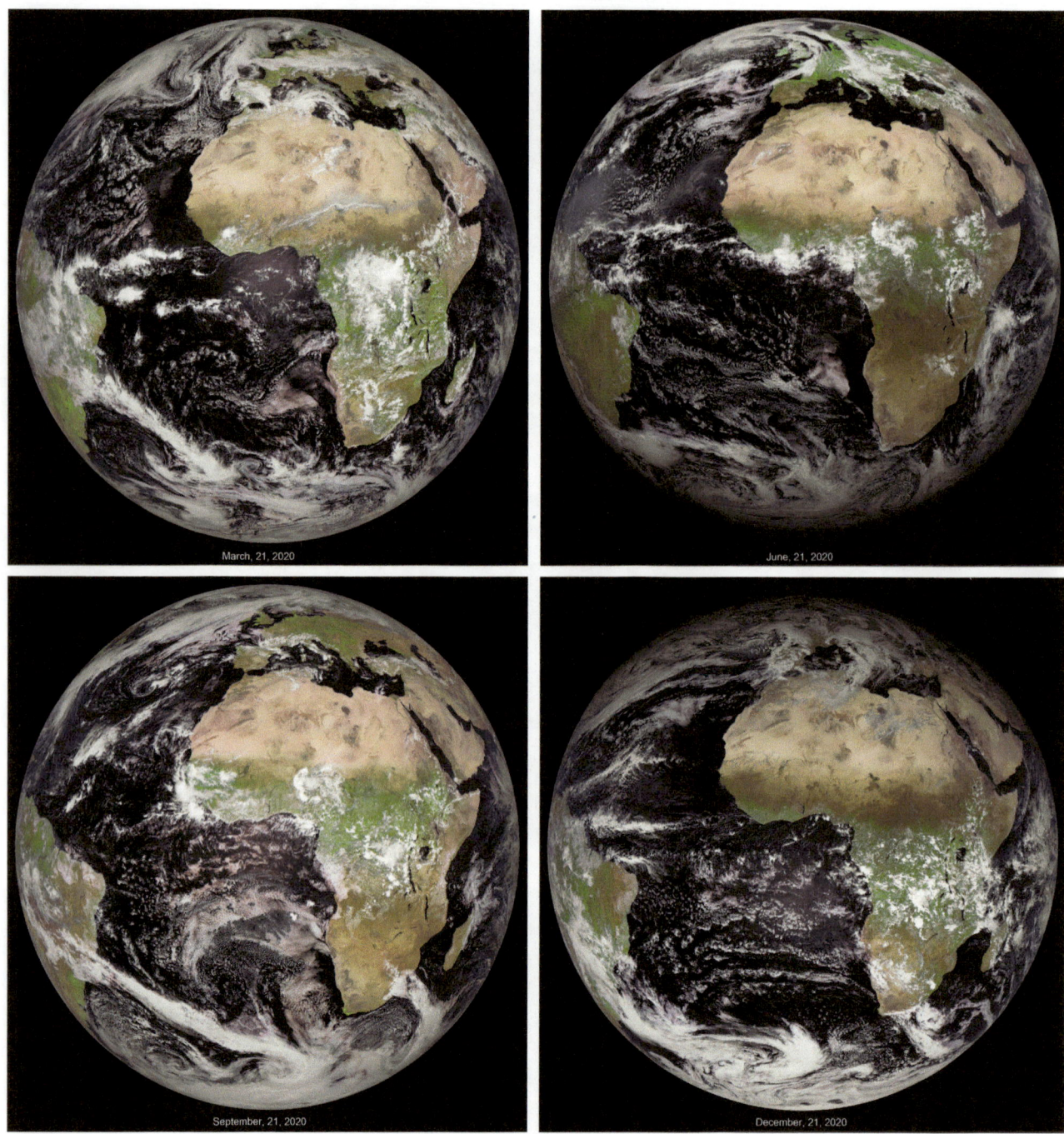

145 Full Earth images at noon also show this seasonal effect. At the beginning of spring (top left) and autumn (bottom left), the Earth is evenly illuminated from the North to the South Pole. At the beginning of summer (top right), the shadow of the polar night appears over Antarctica, while at the beginning of winter (bottom right), it can be seen over the Arctic. Meteosat-11, 2020. (Photo: EUMETSAT)

146 At the summer solstice on June 20, Himawari-8 photographed the Earth at midnight. The thin crescent at the top is not the result of an accidentally rotated 90° Earth image. Due to the tilt of the Earth's axis, the northern hemisphere was oriented towards the Sun. In parts of the Arctic, the Sun no longer set. Therefore, the camera on Himawari was able to photograph the areas on its side in the light of the midnight Sun, even though the Sun was actually hiding directly behind the Earth. Himawari-8, June 20, 2017. (Photo: NOAA, Data: JMA – Japanese Meteorological Agency)

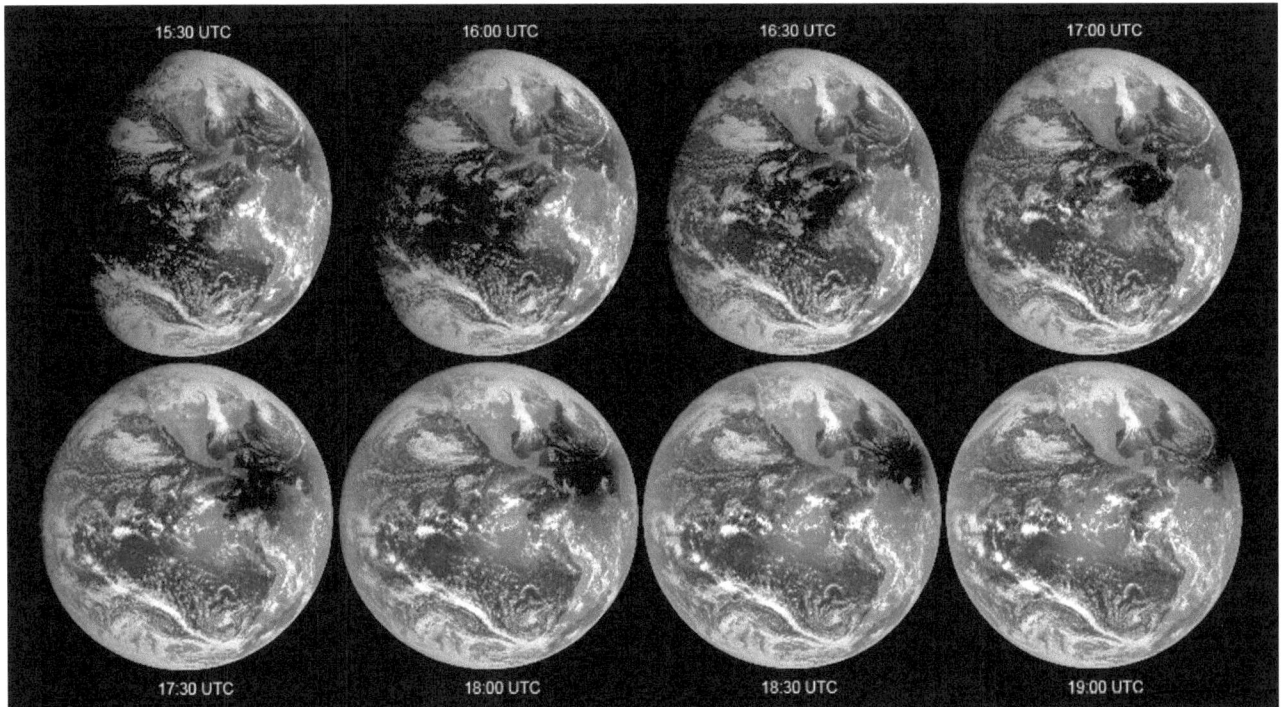

147 The geostationary orbit is an excellent vantage point to track total solar eclipses. The possibility of regular recordings allows tracking the path of the umbra over a longer period of time. In February 1998, GOES-10 showed from its then position how the Moon's umbra appears at the terminator over the Pacific and disappears three and a half hours later at the western edge of the Earth. GOES-10, February 26, 1998. (Photo: NOAA)

148 In the summer of 1999, a total solar eclipse could be observed in Europe, assuming a clear view of the sky. The shadow of the Moon is recognizable as a significantly darkened area. In the excerpt of the Meteosat image from August 11, 1999, you can see the umbra over southern Germany as it projects onto a dense cloud cover. Only occasional gaps in the clouds allowed a view of the rare natural spectacle at that time. Meteosat-7, August 11, 1999. (Photo: EUMETSAT)

149 The image of the solar eclipse from March 2016 at the time of maximum coverage duration comes from the Japanese geostationary satellite Himawari-8. The path of totality of this eclipse moved from Indonesia across the Pacific eastwards. Himawari-8, March 9, 2016. (Photo: NOAA, JMA - Japanese Meteorological Agency)

150 On March 29, 2006, a total solar eclipse occurred, during which the Moon's umbra moved from the east coast of South America to Central Asia. At 10:00 UTC, it stood over the cloudless Sahara, where it stands out clearly in the image from Meteosat-8. Meteosat-8, March 29, 2006. (Photo: Maximilian Reuter, IUP-IFE, University of Bremen, Data: EUMETSAT)

151 The Moon occasionally appears in close proximity to the Earth as photographed by geostationary satellites. If these images show a "full Earth" it is always a full Moon, as the Sun, satellite, Earth, and Moon are then approximately in a line and our companion is also fully illuminated. The enormous difference in brightness and color between the two celestial bodies is clearly visible in the section of an image from Electro-L shown here. The Moon, rendered in grayscale, stands in stark contrast to the colorful Earth. Electro-L, 2013. (Photo: Roscosmos, Vitaliy Egorov, CC BY 3.0)

152 In this image taken by the western GOES satellite, the rising Moon is still half obscured by the Earth. Light refraction in the atmosphere distorts its lower edge. GOES-17, March 6, 2020. (Photo: NOAA)

7. Unmanned Moon Missions - From Cosmic Proximity

The Moon is the only celestial body that we can observe details of with the naked eye from Earth without the use of tools. The opposite perspective – what does the Earth look like from the Moon – has long occupied authors of astronomical depictions of both scientific and utopian nature. Until the beginning of the space age more than 60 years ago, one had to be content with speculations; only after that could this question be answered. The only certainty was the size of the Earth in the lunar sky: Since the Earth is three and a half times larger than the Moon and the Moon has an apparent diameter of 0.5 degrees as seen from Earth, our home planet appears to be about 1.9 degrees large from a lunar perspective.

The space endeavors in the 60s of the last century were characterized by the "race to the Moon" between the United States and the then USSR. Their unmanned automatic probes were primarily used to explore the lunar surface as terrain where the manned landings were to take place. So far in this book, the Earth has always been the object of desire, the focus of the recording devices, but from the Moon's distance, it plays only a subordinate role. Our home planet now only appears randomly in the field of view of the lenses or is used to loosen up an otherwise rigid measurement program.

Cameras were standard equipment on every Moon mission. Due to the lack of atmosphere, the Moon offers an unobstructed view of its surface, which was then only insufficiently known. The smallest structures with an extent of about 1 km that could be detected from the ground were found in a photographic lunar atlas compiled from images of a telescope with a 1.5 m mirror diameter. For the preparation of the manned landings, much more precise information was needed from potential landing sites. After a series of missions, the Ranger missions, which aimed for a successful hard landing and as accurate as possible photographic documentation of the impact site, NASA initiated the Lunar Orbiter program with five launches between August 1966 and August 1967. It aimed to map the Moon very accurately from low orbits. Each orbiter was equipped with a camera and two lenses that together provided a pair of images, consisting of a wide-angle shot and a telephoto shot from its central part. Since the required spatial resolution should only be a few 10 m, the TV cameras then common in satellites in low Earth orbits were not suitable. Instead, the quality of analog film material was needed. For this, an automatic photo lab was devised that developed, fixed, and scanned each exposed shot on board. This information was finally transmitted to Earth, from which the original image could be reconstructed. The black-and-white film material carried had a finite length and was only sufficient for about 210 pairs of images. Unplanned Earth shots had to be well thought out. This was not only due to the limited film available, which served to map potential Apollo landing sites; in order to target the Earth, the entire probe had to be realigned - a risky maneuver in the early days of Moon flight. It took a coincidence and some persuasion from NASA officials until approval was obtained. As Lunar Orbiter-1 orbited the Moon, orbit calculations revealed that on August 23, 1966, just as the probe disappears behind the Moon, the Earth would appear next to the Moon's horizon. It was decided to align Lunar Orbiter-1 accordingly to take the first photo of the Earth, seen from another celestial body. It shows our home planet half illuminated next to the lunar surface and although it was taken from a novel, hitherto unreached perspective, it received little notice. Only in professional circles was its uniqueness recognized. When NASA released the image again later, it was rotated 90 degrees in the press print and that's how we know it today: from a setting behind the Moon to a rising above the Moon.

In total, the five Lunar Orbiters sent back less than a handful of Earth-containing photos. All are taken from lunar distance. The camera could not be used during the flight to the Moon. Once in operation, the film would have had to be advanced with a new exposure at least every 8 hours to prevent the developer from sticking. Upon arrival in lunar orbit, a significant part of the film stock would have

M. Gottwald, *The Earth*,
https://doi.org/10.1007/978-3-662-69633-0_7

already been used up. The more than 40-year-old images of the Lunar Orbiter probes are now being reprocessed in a complex digital process. With the help of new image processing techniques, image errors can be eliminated and the recognizable detail richness in the images can be increased. The photo from August 23, 1966 also shines in new splendor.

Around the same time as the Lunar Orbiter ventures, NASA conducted the Surveyor program. Five probes successfully landed on the lunar surface. During their stay, they examined the lunar soil on site in preparation for manned Moon flights, also using numerous images from the onboard TV cameras. The second successful landing device, Surveyor 3, not only aimed at the ground, but also upwards to the Earth. In April 1967, we thus witnessed a solar eclipse on the Moon for the first time. With an Earth shadow three and a half times larger, the eclipse lasted significantly longer than a comparable event on Earth. The photo turned out relatively blurry, but clearly shows how the light of the Sun standing behind the Earth is refracted in the atmosphere and forms an incomplete ring. A few days later, the Earth was again the target of Surveyor 3's camera. Using filters, the first color photo of our planet from the Moon was created. In total, the Surveyor landers delivered some 10,000 images; only a small part of them has been published so far. Similar to the images of the Lunar Orbiter missions, the Surveyor photos stored on 70 mm film are currently undergoing a digitization process.

What did similar results of Russian space flights look like? The restrained information policy of the Russian side made it difficult at the time to keep track of the results of their automatic probes. Although the USSR's space program achieved the first soft landing on the Moon, the first operation of a probe in a lunar orbit, the first unmanned transport of Moon rock to Earth, and the first tour of a robot over the lunar surface, no images of the Earth are known from any of these missions. Such do exist, however, from the Zond program, which served as a test phase for manned Russian missions. The orbit only led around the Moon and then back to Earth. Between September 1968 and October 1970, four flights, Zond-5 to Zond-8, were able to be carried out on such orbits. Each delivered a few images, with Zond-7 even in color. Although

the quality of most of these photos was quite high and Zond-7 even transmitted an image of the full Earth for the first time while the NASA probes only ever showed our home partially illuminated, they aroused little public interest. They simply came too late. NASA's manned Apollo program had already provided excellent images of the Earth from lunar distances with medium format cameras from December 1968. The Zond images did not even come close to these. With the last landing of the Russian automatic laboratory Luna 24 and the return of a capsule with Moon rock, the first phase of Moon exploration ended in August 1976. It was dominated by the question of who would reach our satellite first.

For the next nearly 20 years, the Moon had only low priority in space travel. It was not until the beginning of the 90s of the 20th century that it came back into focus. Particularly the success of Clementine, a joint project of NASA and the American Department of Defense, heralded 22 years after NASA's last Moon visit with Apollo 17 in December 1972 a reconnaissance of lunar space flights. The photographic imaging of the Moon's surface was one of the main tasks during Clementine's stay in lunar orbit between February and May 1994. Here also the first photo of the Earth was taken again, showing it as a colorful home of life against the black sky background above the inhospitably gray lunar surface. In fact, this well-known image is a photomontage - the Earth was actually much higher and was placed lower in the image to enhance the effect.

After Clementine, other space nations entered the Moon stage. This included Europe, whose space agency ESA launched SMART-1 to the Moon in September 2003. The propulsion of this probe was a novelty in the history of lunar missions. An ion engine sent SMART-1 on a spiral path to the Earth's satellite. The usual way to our companion was a direct one, starting with a few minutes of ignition of a chemical motor from an Earth orbit, which propelled the probe towards the Moon. It then moved on a transfer orbit away from the Earth's gravitational field towards the Moon, slowing down on the way there due to the Earth's gravity until it reached the Moon's gravitational field and was accelerated again. Once there, it either only performed a fly-by and was steered back towards Earth around the Moon or it swung into an orbit and sometimes even deployed a lander from it. In an ion engine, electrically charged particles are accelerated and ejected in the rocket engine. This creates a thrust that is much weaker than that of a conventional engine. While the chemical engine as an explosive engine only works for a few minutes before its fuel supply is exhausted, an ion engine can be operated continuously with significantly lower consumption of noble gas fuel. Every rocket engine operation is done to create a change in speed. According to the laws of gravity, this is equivalent to a change in orbit. Therefore, a constantly working ion engine produces a constant change in speed, leading to an ever-widening orbit. After the launch and reaching the geostationary transfer orbit, the ion engine of SMART-1 ignited, so that the probe slowly but continuously moved away from the Earth towards the Moon. In November 2004, SMART-1 finally reached the lunar orbit, from which it completed its measurement program until September 2006 to finally be deliberately crashed onto the lunar surface. On its way to the Moon, shortly before its arrival, SMART-1 experienced a total lunar eclipse in October 2004. SMART-1 looked at the Earth in the foreground and the Moon passing behind it with the Sun as a light source in the background. As soon as the Moon reaches the umbra shadow of the Earth, it disappears in the recorded image sequence only to emerge from it again later.

The next visit to the Moon took place in September 2007 by Japan with the Moon probe Kaguya. It had a special feature, a High Definition camera system (HDTV). In addition to the scientific tasks, it was supposed to transmit an image of the Moon for the public, primarily from the perspective of a virtual Astronaut. The HDTV system consisted of a wide-angle and a telephoto lens, each oriented in opposite directions. Activated for the first time on the way to the Moon, it delivered images of the Earth of outstanding quality. From the Moon's orbit, where Kaguya remained until June 2009, the video sequences with the rising and setting Earth and the Moon's surface in the foreground were particularly impressive.

They reminded of similar scenes from the time of the manned Apollo Moon flights.

China, India, South Korea, and even Israel have now joined the ranks of nations with Moon missions. Particularly interesting was China's Chang'e-3 in December 2013. The namesake was the Chinese Moon goddess. Chang'e-3 was not an orbiter, but deployed a lander with the mobile robot Yutu, named after the companion of the Moon goddess, the Jade Rabbit, on the Moon's surface. Yutu covered about 100 m on the Moon's surface. With its camera, it managed to capture the Earth. The scenery resembles that of Surveyor 3, but with significantly better resolution. A little later, in October 2014, Chang'e 5-T1 followed as preparation for the Chang'e 5 mission, which has since successfully brought Moon rock back to Earth. Chang'e 5-T1 orbited the Moon and then returned to Earth. Its images show interesting constellations of the Earth and Moon system.

Of course, NASA also continued its Moon program. Many of its Moon probes, however, no longer aimed at the most detailed mapping of the Earth's satellite, but dealt with questions of the structure of the Moon's interior or the composition of its surface. The highest resolution images of the Moon to date are provided by NASA's "Moon explorer", the Lunar Reconnaissance Orbiter (LRO) since June 2009. LRO orbits the Moon in an orbit that takes the probe only 50 km above the surface of our satellite at the lowest part of the orbit, resulting in images that can show objects of only 50 cm in size. This made it possible to photograph the objects made by human hands and landed on the Moon during the first phase of Moon exploration in the 60s and 70s of the last century. LRO is also ideal for a detailed look back at the Earth. It appears in the LRO images from almost 400,000 km away with a clarity that reminds of images from the geostationary orbit from the previous chapter. One of these photos, taken in May 2014, makes it clear that we are not alone in the solar system. Far behind the "full Earth" you can see our neighbor Mars faintly shining in the dark sky. The high resolution makes it appear not just as a point of light, but even as a small disc.

153 The first view of the Earth from the distance of another celestial body (above), taken by Lunar Orbiter 1. On August 23, 1966, the half-illuminated Earth appeared next to the Moon's horizon, shortly before it set for the probe. Here you see the later published version, in which the orientation was rotated by 90° to represent a seemingly rising Earth. Numerous artifacts affect the image. At the Lunar and Planetary Institute in Houston, Texas, there is a complete catalog of Lunar Orbiter images. All photos were processed using modern image processing methods and visible artifacts were largely corrected. This led to the cleaned image shown below. Lunar Orbiter 1, August 23, 1966. (Photos: NASA, LPI)

154 In 2007, the Lunar Orbiter Image Recovery Project (LOIRP) began with the participation of NASA. The goal was to digitize the original images from Lunar Orbiter and process them in high quality. From the first image of the Earth from the vicinity of the Moon (Figure 153), this cleaned, significantly better image of the Earth and the Moon's horizon was created. Lunar Orbiter 1, August 23, 1966. (Photos: NASA, LPI, LOIRP)

155 While the previous illustrations show the high-resolution photo of the telephoto lens, the wide-angle lens provided a much larger section of the lunar surface; to the left, the Ziolkowski crater with its dark crater floor. This image clarifies the correct orientation at the time of the recording with the Earth standing near the lunar horizon near its equator. Lunar Orbiter 1, August 23, 1966. (Photo: NASA, LPI)

156 Two days after the first Earth photo was taken from lunar distance, the opportunity arose again to capture our home planet at about the same place above the horizon, this time with a slightly smaller Earth phase. The wide-angle shot (above) shows roughly the same lunar region as the earlier version. Also as part of the LOIRP activities, the digitized high-resolution version of this scene was created (below). It consists of two adjacent individual images. Lunar Orbiter 1, August 25, 1966. (Photos: NASA, LPI, LOIRP)

157 An approximate full Earth, taken by Lunar Orbiter 5. The view goes towards the Indian Ocean. Clearly visible on the left is cloudless Africa, the Arabian Peninsula, and the eastern Mediterranean. Lunar Orbiter 5, August 8, 1967. (Photo: NASA, LPI)

158 The first color photo of the Earth from the surface of the Moon, taken by Surveyor 3. The camera looked through filters at South America during the recording, which was separated into day and night by the terminator. Surveyor 3, April 30, 1967. (Photo: NASA)

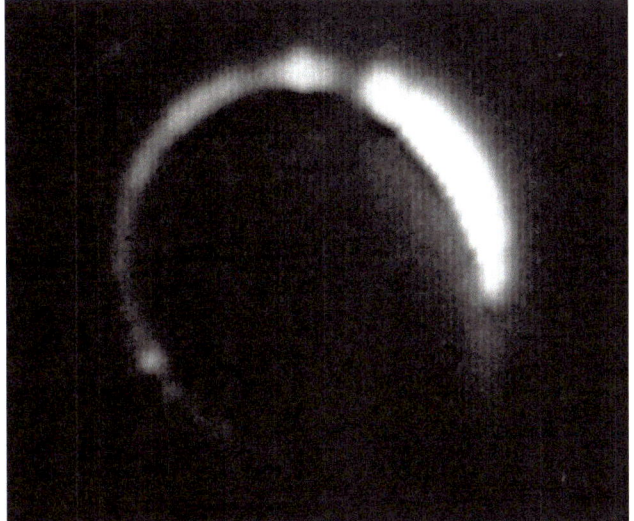

159 Among the Earth shots that Surveyor 3 had taken from the lunar surface, there is also a short sequence of the first total solar eclipse observed from this perspective. The Earth completely covers the Sun due to its apparent size. In the Earth's atmosphere, the Sunlight is refracted and forms a bright ring around the Earth; interrupted in some places by clouds. Surveyor 3, April 24, 1967. (Photo: NASA)

160 At the Lunar and Planetary Laboratory of the University of Arizona, the Surveyor images available on 70 mm are digitally processed and archived. Here the photo shows an image of the half-Earth from Surveyor 7 from early 1968. Surveyor 7, January/February 1968. (Photo: NASA, JPL, University of Arizona Lunar and Planetary Laboratory)

161 The Russian probe Zond 5 was 90000 km on its way to the Moon when it managed to take this Earth shot. A later defect prevented its photo campaign during the lunar orbit. Zond 5, September 16, 1968. (Photo: Don P. Mitchell, Data: Moscow State University for Geodesy and Cartography)

162 Upon its return from the Moon, the Zond 6 probe crashed on Earth. Although the film canister was damaged in the process, several shots were saved, including this one with the Earth standing above the lunar horizon. Zond 6, November 14, 1968. (Photo: Don P. Mitchell, Data: Moscow State University for Geodesy and Cartography)

163 A sequence of the Earth's setting, compiled from 3 shots of the Zond 7 probe. The second image in it is a simulation created from the other images to be able to continuously represent the setting. Zond 7, August 9, 1969. (Photo: Ted Stryk, Data: Moscow State University for Geodesy and Cartography)

165 This color image of the Earth standing above the horizon was created two days later during its lunar orbit. Shortly after the first manned Moon landing, this image had to compete with the results of Apollo 11. Zond 7, August 11, 1969. (Photo: Don P. Mitchell, Data: Moscow State University for Geodesy and Cartography)

164 On its journey to the Moon, Zond 7 photographed the Earth in color. Zond 7, August 9, 1969. (Photo: Don P. Mitchell, Data: Moscow State University for Geodesy and Cartography)

166 The last probe in the Zond series, Zond 8, visited the Moon in October 1970. Images of the Earth setting on the lunar horizon had become standard repertoire for such ventures. The overview and detailed images from Zond 8 were of very good quality. Zond 8, October 24, 1970. (Photo: Don P. Mitchell, Data: Moscow State University for Geodesy and Cartography)

168 The Earth from a distance of 75000 km, captured by Clementine on her way to the Moon. India and the tip of Saudi Arabia can be seen on the horizon. Clementine, January 1994. (Photo: NASA, JPL, USGS)

169 A mosaic of the Earth from Moon distance, composed of numerous individual images. Clementine, April 11, 1994. (Photo: NASA, JPL, USGS)

167 Above: Clementine's view of the Earth standing above the Moon. The Earth was brought a little closer to the Moon in image processing and enhanced in color. For the first time since the end of the Apollo missions, such a photo could be seen again. Clementine, April 1994. (Photo: NASA, JPL, USGS)

170 On its way from Earth to the Moon with an ion engine, the European probe SMART-1 followed a complex trajectory, which also took the probe far from the Moon. At the time of the total lunar eclipse in October 2004, it was 660000 km from the Moon and 290000 km from the Earth. The sequence of Moon shots shows the course of the eclipse from left to right. The two Earth hemispheres were taken shortly before and after the eclipse and convey the correct size ratios from SMART's position. SMART-1, October 28, 2004. (Photo: ESA/Space-X)

171 After the successful launch of Kaguya, this image of the Earth was created from a distance of 110000 km. Kaguya, September 29, 2007. (Photo: JAXA, NHK)

172 In the HDTV telephoto shot of an Earthrise (above) from April 2008, you can see the upside-down Earth above the lunar surface near the Moon's south pole on the far side. We are looking at the Pacific and see parts of North America on the left on Earth. The lower image shows a Sunset in November 2007, also over its south pole with an upside-down Earth, on which you can see Australia on the left and parts of Asia on the right. Kaguya, April 6, 2008 and November 7, 2007. (Photos: JAXA, NHK)

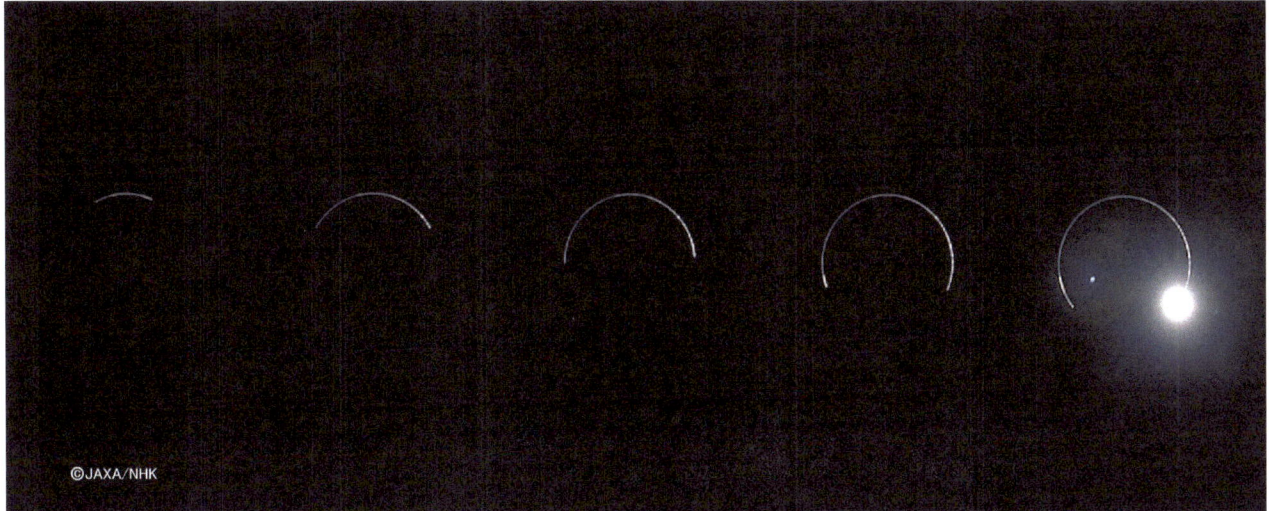

173 Now sequences of an Earthrise captured with the HDTV camera in April 2008 (above) and a Sunset in November 2007 (below). Kaguya, April 6, 2008 and November 7, 2007. (Photos: JAXA, NHK)

174 In February 2009, Kaguya witnessed a penumbral lunar eclipse as seen from Earth. From the Moon's perspective, this appeared as a solar eclipse. The HDTV camera looked at the nearby dark lunar horizon, over which the Sun covered by the Earth rises in the image sequence. The Earth's atmosphere appears as a thin glowing segment, which becomes more and more of a ring, the higher the Earth. and the Sun rise. The last image ends the complete coverage of the Sun by the Earth. Kaguya, February 10, 2009. (Photo: JAXA, NHK)

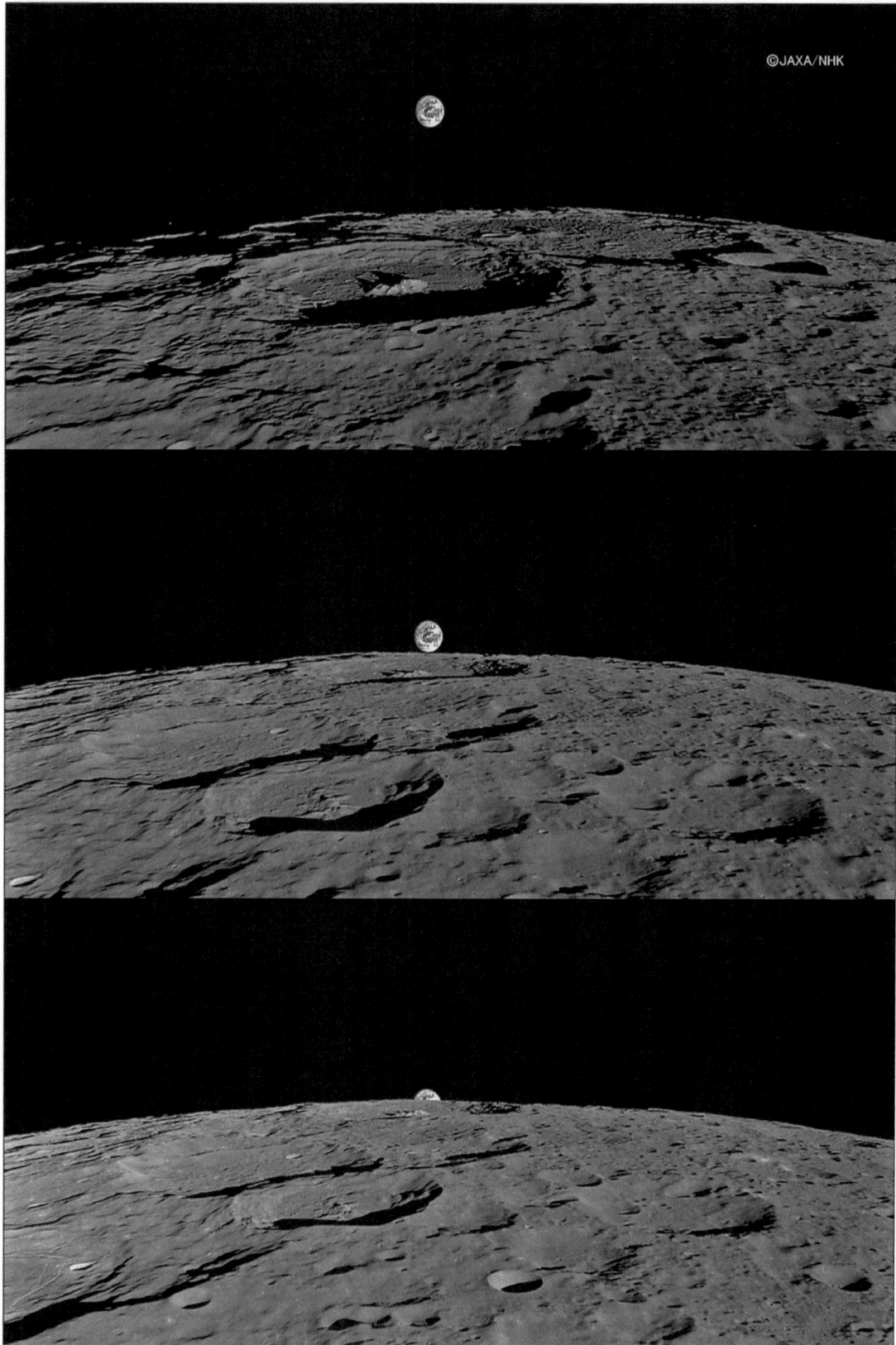

©JAXA/NHK

175 A few weeks after reaching the Moon, Kaguya pointed its wide-angle camera at the rising Earth. The sequence illustrates the difference between the lifeless gray world of the Moon and the colorfulness of the inhabited Earth. In the top image, the Earth is exactly above the Plaskett crater at the North Pole on the far side of the Moon. Kaguya, November 11, 2007. (Photo: JAXA, NHK)

176 Left: After launch, when already 70,000 km on the way to the Moon had been covered, India's first Moon probe had the half-Earth in view. The only landmass visible is Australia's northwest coast, darkly delineated from the Sunlight reflecting Indian Ocean. Chandrayaan-1, October 29, 2008. (Photo: ISRO)

177 Below: The Chinese Moon probe Chang'e 5-T1 was sent on its way to test several Moon flight maneuvers including return to Earth. After orbiting the Moon, Chang'e 5-T1 first returned to Earth. This view of our home planet was taken at the end of October. The constellation with a view of Australia resembles that of Chandrayaan-1 on this page. Chang'e 5- T1, October 2014. (Photo: CNSA/CLEP)

178 One year before Chang'e 5-T1, its predecessor Chang'e 3 successfully deployed the Yutu lander on the Moon's surface on December 14, 2013. Around Christmas time, it observed from there the waxing Earth in the Moon's sky. Chang'e 3, December 20-24, 2013. (Photo: CNSA/CLEP)

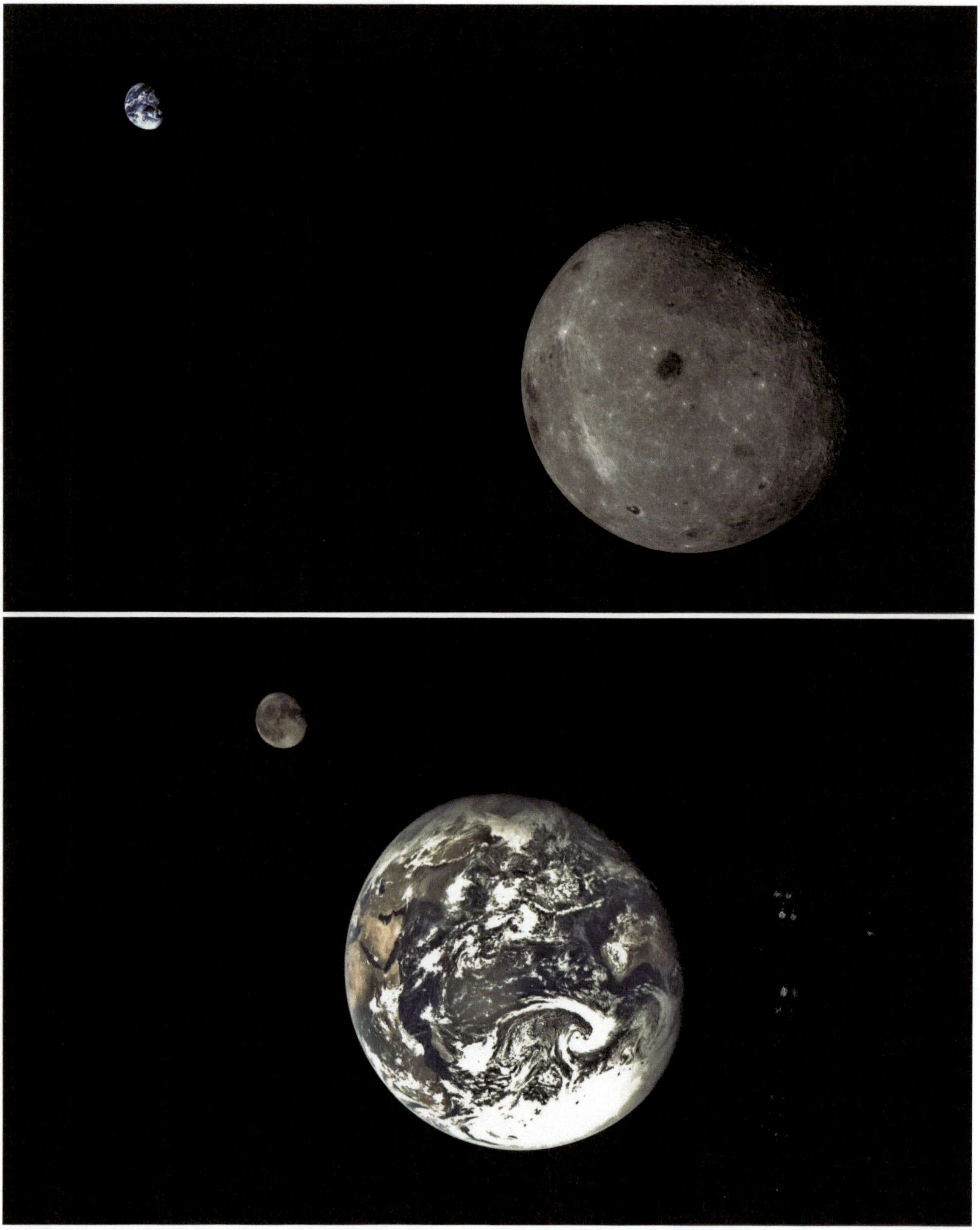

179 On its flight, Chang'e 5-T1 was able to capture the Earth-Moon system from two sides. Once the probe was closer to the Moon than the Earth (above). The dark structure in the middle of the far side of the Moon is the Mare Moscoviense. In the continuation of the mission, it was 540,000 km away from the Earth, while the distance to the Moon in the background was 920,000 km (below). Chang'e 5-T1, October 28, 2014 and November 9, 2014. (Photo: CNSA/CLEP)

180 The highest resolution from Moon distance is achieved by the telecamera of the Lunar Reconnaissance Orbiter. It only delivers black and white images. However, the digitally available images can be arbitrarily manipulated in the computer. If they are colored as naturally as possible, one obtains images of the Earth that appear as if they were obtained from geostationary orbit. LRO, August 9, 2010. (Photo: NASA, GSFC, Arizona State University, Gordan Ugarkovic)

181 In addition to the telephoto lens, LRO also has a low-resolution wide-angle camera. It captures in different wavelength ranges, from which a color composite can be created. One such shows the rising Earth in February 2014. While the Moon appears in its gray tones, our blue planet stands as a colorful contrast just above the horizon. LRO, February 1, 2014. (Photo: NASA, GSFC, Arizona State University)

182 If you combine the image information of the telephoto and wide-angle camera, you get high-resolution color photos. The wide-angle optics provided the color, the telephoto camera on the other hand a high-resolution grayscale image. From both, this unique image of our home planet as a "Blue Marble" was created when the Lunar Reconnaissance Orbiter orbited the Moon at a height of 134 km. LRO, October 12, 2015. (Photo: NASA, GSFC, Arizona State University)

183 In May 2012, over the North Pacific, an annular solar eclipse took place. LRO captured the slightly brighter shadow of the Moon moving towards the west coast of North America (above). Five years later, in August 2017, residents in parts of the United States were able to observe a total solar eclipse (below). The LRO control team saw it in Arizona, which was not covered by the umbra shadow, from lunar distance. LRO, May 20, 2012 and August 21, 2017. (Photo: NASA, GSFC, Arizona State University)

184 In May 2014, LRO managed to show two planets of our solar system from the Moon's orbit. The mosaic composed of two images shows the Earth at a distance of almost 380,000 km and the 112.5 million km distant Mars as a faint small disc above it. The Mars image was enhanced for better visibility. LRO, May 24, 2014. (Photo: NASA, GSFC, Arizona State University)

8. Manned Moon Missions - The Men on the Moon

Of the 3.8 billion people who populated the Earth around 1970, 27 had the privilege of truly seeing our home for the first time from a great distance as part of the cosmos. They all belonged to NASA's manned Moon program. Nine 3-man crews visited our companion between December 1968 and December 1972 with the Apollo capsules. There were two planned stays in lunar orbit, an unplanned circumnavigation of the Moon, which brought the stricken Apollo 13 mission back to Earth, and six successful Moon landings.

Each Apollo mission was equipped with photo, film and TV cameras, which were supposed to accurately document the Astronaut's journey from Earth to the Moon and back, their stay in orbit around the Moon and on its surface. The equipment for analog film recordings was high-quality medium format cameras as workhorse, later supplemented by 35mm cameras, always equipped with very good lenses. This ensured that excellent recordings were brought to Earth, far better than what could be achieved simultaneously with the unmanned probes of the previous chapter. These photos served not only as scientific imagery but were also intended to radiate special public effectiveness. It was as it is still today with flights in low Earth orbits: While automatic probes provide us with data from which images of the Earth with high scientific use can be created, astronauts see their home planet with different eyes. They can accordingly select impressive scenes and objects when looking out of their space station and capture them on film or memory card. The camera of an unmanned probe has no feeling for the beauty of the sight when it is directed at the Earth – an astronaut does. Only he can capture this and share it with us.

For the first time, three people with Apollo 8 were sent towards the Moon with a Saturn V rocket shortly before Christmas 1968. The mission of the crew with Frank Borman, James Lovell and Bill Anders was to orbit the Moon ten times and then return safely to Earth. Already three and a half hours after the start, images were created which show almost the entire Earth and shortly afterwards, from a distance of 30,000 km, the first image of a full Earth exists, a blue-white view of the Atlantic with parts of South America and Africa. Attempts to offer a similar perspective to the viewers sitting worldwide at their televisions during the transit phase to the Moon were only successful near the Moon, when the on-board TV camera delivered a black-and-white image of the Earth, on which mainly cloud fields could be identified. It was not until the late afternoon of December 24 that images were taken that still count among the icons of space photography today. Apollo 8 was in its fourth lunar orbit, just coming out from behind the Moon's edge near the equator, and performed a maneuver that caused the spacecraft to rotate around its longitudinal axis. Bill Anders observed and photographed the lunar surface through one of the windows with a camera containing black and white film when he saw the Earth rise over the horizon of our satellite. His first shot with the Earth standing directly above the Moon was taken - however, without reproducing the color of our home planet. Until a magazine with color film was found and loaded, about a minute passed. By then, the Earth had wandered out of the window due to the rolling motion of Apollo 8, but appeared shortly thereafter in the next. Now the picture was taken, which for some is considered one of the most important photos of the last century and stands as a symbol for mankind's advance into space, "Earthrise - Earthrise over the Moon" shortly thereafter another one before the Earth disappeared from view again. Both shots illustrate what James Lovell said after the flight of Apollo 8 about the view of our home: "The Earth from here is a grand oasis in the big vastness of space. The special feature of the first Earthrise observed by humans in a foreign world was also enhanced by the Christmas message sent from the spaceship to Earth on the same day on Christmas Eve in the form of the first sentences of the biblical genesis. Why didn't Apollo 8 see the Earth in the earlier

Moon orbits? It was simply the coincidence that exactly at the time when the Earth appeared on the horizon of the Moon due to the movement of the spacecraft around the Moon, a roll maneuver was also executed and it was thus visible at all through one of the windows from inside the Apollo capsule. In later Moon orbits, it was possible to capture further Earthrises. The surprising and magical moment of the first time, however, did not occur again. The rise of the Earth over the lunar horizon in the fourth lunar orbit is documented in only three photos. For the 45th anniversary of the Apollo 8 mission, NASA created a video that realistically captured these moments. From the data and photos of Apollo 8, its path was well known, so that based on data from the Lunar Reconnaissance Orbiter (see previous chapter) it could be simulated how the lunar surface appeared to the astronauts when looking out of the windows of the Apollo capsule. The sight of the Earth was recreated from data of a weather satellite at that time, which reflected the cloud conditions and a modern Earth observation satellite, from which an accurate image of the Earth's surface could be derived. Accompanied by the audio recording, we become witnesses in the video, so to speak, as Apollo 8 moves over the lunar surface, the Earth rises on the horizon and the astronauts are overwhelmed by the sight, yet they manage to capture this moment.

The subsequent Apollo flights carried out scientifically and technically more complex tasks, with Apollo 9 staying only in Earth orbit, to test the lunar module among other things. A little later on Apollo 10, two astronauts approached the lunar surface with the lunar module for the first time to within a few kilometers while Apollo 11 finally achieved the first manned Moon landing. All further missions up to Apollo 17, apart from Apollo 13, completed increasingly extended stays on the Moon. Each of these Apollo crews brought back a wealth of photographic material. While Apollo 8 had 865 shots on the exposed medium format films, this number had already risen to 1408 for Apollo 11 and the last mission Apollo 17 finally contributed 3581 photos. If the Earth was visible on them, it continued to play the role of a livable place against the backdrop of the dark universe and the gray inhospitable lunar surface. Of course, each Apollo crew continued to document how the Earth and the Moon presented themselves on the

journey there and back. Apollo 12 even managed to take a photo of a solar eclipse on its return, in which the Earth obscured the Sun. Since due to the geometry the Earth naturally appeared much larger than the disk of the Sun because of its proximity, the phenomenon of the solar corona known from terrestrially observed eclipses could not be seen. It was impressive, however, how the Sunlight was scattered in the Earth's atmosphere and formed a wide circulating narrow bright ring - just as the unmanned probes Surveyor 3 and Kaguya had also seen in the previous chapter.

Earthrises from the command capsule orbiting the Moon continued to be part of the repertoire. However, the experience of Apollo 8 now also created scenes that could be designed more consciously. Outstanding here is the sequence from Apollo 11, in which it was captured how the Earth step by step appears on the Moon horizon and rises. Such shots were possible when the Apollo capsules in their orbit switched from the back of the Moon to the front visible to us. This was an important moment in the flight plan, because then contact between Earth and spacecraft could be reestablished. Since the photos taken at this time were made from the elevated position of the orbit, the Earth could already be spotted while the astronauts were still over edge-near regions of the far side of the Moon.

With the use of the lunar module, a new Earth perspective was created: Either the Earth shone above the lander standing on the Moon, peeked through one of its windows or appeared just then in the Moon sky, as the lunar module after successful Moon stay returned to the command unit in the Moon orbit. This scene was accidentally captured by Apollo 11. Michael Collins, the astronaut left behind in the capsule, followed the approach of the landing gear Eagle with his colleagues Neil Armstrong and Buzz Aldrin on board. Exactly at the right moment, shortly before docking, Eagle formed in front of the Moon background with the Earth standing above an impressive constellation. Later, Collins described how he became aware that in this picture 3 billion Earth inhabitants, two explorers and a Moon can be seen - while he as a photographer discreetly stayed in the background. None of the Earth shots taken by the astronauts from Moon distance showed a full Earth. All manned landings on the Moon had its illuminated front side as a target. Therefore, Earth, Moon and Sun always formed a triangular constellation with lateral illumination. The longer the stay on the Moon lasted, the more the phase shapes of the Earth in the Apollo shots could differ. This is already evident in the two mentioned Apollo 11 images. When the astronauts of the first Moon landing arrived at the Moon on July 19, its phase from the Earth's point of view was just between new Moon and first quarter, the Earth on the other hand appeared between first quarter and full Earth. When the lander returned two days later, this constellation had already noticeably changed; a larger part of the Moon was now in Sunlight while on Earth the illuminated portion had correspondingly decreased. However, there is a shot in which the Sun stood on the entire observable Earth hemisphere above the horizon. It was taken by Apollo 17 on December 7, 1972 after the launch on the way to the Moon from a distance of 29000 km. This corresponded approximately to the perspective of a geostationary satellite, only that the Apollo capsule was not above the equator, but was just crossing southern latitudes and enjoyed an unobstructed view of Antarctica. The photo was known under the designation "Blue Marble" and is considered one of the most outstanding shots that NASA ever published. At the time of its creation, it was first recognized that we must understand the Earth, our home, as a complex system with only finite resources. The environmental movements were born. NASA's Blue Marble then illustrated like hardly any other picture, the beauty of the Earth, but also its vulnerability and thus achieved cult status. Currently, NASA is preparing again for the sending of humans to the Moon. As part of the Artemis program with international participation, visitors from Earth are expected to land there no later than about 55 years after Apollo 17 last had astronauts set foot on the Moon. On November 16, 2022, Artemis-1 launched as an unmanned Orion spacecraft together with a propulsion and supply unit. Artemis-1 orbited the Moon for several days, sometimes at a great distance. from 60,000 km and then set off again for Earth. This mission produced images as we know them from the Apollo flights, but sometimes from a new, unfamiliar perspective.

185 Shortly after the shot towards the Moon, the crew of Apollo 8 took the first photos of the Earth, here Florida and the Bahamas from a height of 6000 km. Apollo 8, December 1968. (Photo: Earth Science and Remote Sensing Unit, NASA Johnson Space Center, AS08-16-2581)

186 At a distance of about 30,000 km, Apollo 8 made the first human-made image of a (almost) full Earth. Apollo 8, December 1968. (Photo: Earth Science and Remote Sensing Unit, NASA Johnson Space Center, AS08-16-2593)

187 When Apollo 8 was already gravitationally bound by the Moon, viewers sitting in front of the television on Earth could also take a look at their planet. The transmission from a distance of 325,000 km captured an Earth where mainly cloud cover was visible in black and white mode. Apollo 8, December 1968. (Photo: NASA)

188 The first photo of the rising Earth above the lunar horizon. During the attitude maneuver, the Earth appeared surprisingly in one of the windows of the Apollo capsule. The image is only in black and white, as the camera accessible to Bill Anders was loaded with such a film. As with many unmanned probes, the image was rotated 90° upon publication to show a horizontal lunar horizon. In fact, the scene took place near the lunar equator and its edge ran vertically through the image. Apollo 8, December 1968. (Photo: Earth Science and Remote Sensing Unit, NASA Johnson Space Center, AS08-13-2329)

189 By the time a magazine with color film was at hand, the Earth had moved out of the window due to the movement of the spacecraft and appeared in the adjacent one. Now the first color image (below) was created, which brought it worldwide fame. A second followed shortly afterwards with a slightly higher standing Earth (above). Apollo 8, December 1968. (Photos: Earth Science and Remote Sensing Unit, NASA Johnson Space Center, AS08-14- 2383/2384)

190 The scenery of the Earthrise captured by Frank Borman in the 7th lunar orbit conveys the impression that the astronauts experienced when looking out of the window of the Apollo capsule. Apollo 8, December 1968. (Photo: Earth Science and Remote Sensing Unit, NASA Johnson Space Center, AS08-14-2392)

191 On the way to the Moon, the Earth appeared to the astronauts of Apollo 10 when looking at the west coast of North America, taken with the normal focal length, initially still filling the image (above). Later, during the transfer phase to the Moon, a telephoto lens was already needed to recognize similar details on Earth, now North Africa. Apollo 10, May 1969. (Photos: Earth Science and Remote Sensing Unit, NASA Johnson Space Center, AS10-34- 5014/5027)

192 For the first time, a photo of the Earth was taken from the lunar module, as two Apollo 10 astronauts approached the lunar surface for testing up to about 15 km. The earthrise took place over the Mare Smythii. Apollo 10, May 1969. (Photos: Earth Science and Remote Sensing Unit, NASA Johnson Space Center, AS10-27-3887/3890)

193 After Apollo 11 was sent on its way to the Moon by igniting the 3rd stage of the Saturn V rocket, this image of parts of North and Central America was taken at a distance of about 10,000 km (above).
Eight days later, the astronauts returned to Earth and took their home planet from a distance of 63000 km with an 80 mm standard lens. (below). The western Indian Ocean reflects the sunlight, south is pointing upwards. Apollo 11, July 1969. (Photos: Earth Science and Remote Sensing Unit, NASA Johnson Space Center, AS11-36-5308/6676)

194 Until their arrival on the Moon on July 19, the Apollo 11 astronauts repeatedly took photos of the slowly shrinking Earth. With the telephoto lens of the Hasselblad camera, numerous details were still recognizable from about 91000 km (top left). On the further way to the Moon at distances of 181000 km (top right), 306000 km (bottom left) and finally shortly before reaching the Moon (bottom right) the Earth appeared significantly smaller in the same lens. Apollo 11, July 1969. (Photos: Earth Science and Remote Sensing Unit, NASA Johnson Space Center, AS11-36- 5340/5353/5378/5402)

195 While Neil Armstrong and Buzz Aldrin were on the Moon's surface, Michael Collins orbited our satellite in the command module. From there he observed, again over the Mare Smythii, how the Earth appeared on the Moon's horizon and then rose slowly due to the movement of the spacecraft. Apollo 11, July 1969. (Photos: Earth Science and Remote Sensing Unit, NASA Johnson Space Center, AS11-44-6547/6548/6549/6551/6557/6564)

197 After the docking of the lunar module, one of the last photos was taken from the Moon orbit before Apollo 11 returned to Earth again. The bright crater in the foreground is Al-Kwharazimi on the far side of the Moon, behind it the dark Mare Smythii reappears. Apollo 11, July 1969. (Photo: Earth Science and Remote Sensing Unit, NASA Johnson Space Center, AS11-44-6645)

196 Left: One of the most famous Earth depictions from Apollo 11 was taken from the command capsule during the return of the lunar module. In a sequence, you can see in the foreground the approaching lunar module while the Earth stands on the horizon and rises higher during the docking process. Apollo 11, July 1969. (Photos: Earth Science and Remote Sensing Unit, NASA Johnson Space Center, AS11-44-6633/6634/6642)

198 The narrow Earth crescent above the Moon's horizon. Apollo 12 was over the crater Pasteur at the time of the shot, the western part of which can be seen in the foreground. Apollo 12, November 1969. (Photo: Earth Science and Remote Sensing Unit, NASA Johnson Space Center, AS12-47-6871)

199 On the return flight from Apollo 12 from the Moon, the Earth covered the Sun. The apparently much smaller Sun disappeared completely behind the Earth, the sunlight scattered in the atmosphere formed a glowing ring. Apollo 12, November 1969. (Photo: Earth Science and Remote Sensing Unit, NASA Johnson Space Center, AS12-53-7925)

201 Apollo 13 narrowly avoided a catastrophe after the explosion of an oxygen tank on the approach to the Moon. Shortly after the injection towards the Moon, everything was still in order and the crew could also devote themselves to photographic tasks like their predecessors. Here you can see the western Pacific area, taken 9 hours after the launch on the way to the Moon. Apollo 13, April 1970. (Photo: Earth Science and Remote Sensing Unit, NASA Johnson Space Center, AS13-60-8593)

200 Left: From the lunar module, the two Apollo 12 astronauts, Charles Conrad and Alan Bean spotted the Earth with the crater Meitner in the foreground shortly before the descent to the surface of our satellite. The sequence shows the ever-rising narrow Earth crescent. Apollo 12, November 1969. (Photos: Earth Science and Remote Sensing Unit, NASA Johnson Space Center, AS12-47- 6879/6889/6895)

202 The landing site of Apollo 14, like that of Apollo 12, was in the equatorial region of the Moon, only 10° away from it. Therefore, both spacecraft moved in similar orbits and when the lunar module separated from the command module, a similar constellation as with Apollo 12 resulted. The Earthrise sequence, taken at exactly this moment, reminds of Apollo 12. The prominent crater in the center of the image is again Meitner. Apollo 14, February 1971. (Photos: Earth Science and Remote Sensing Unit, NASA Johnson Space Center, AS14- 66-9224/9225/9228)

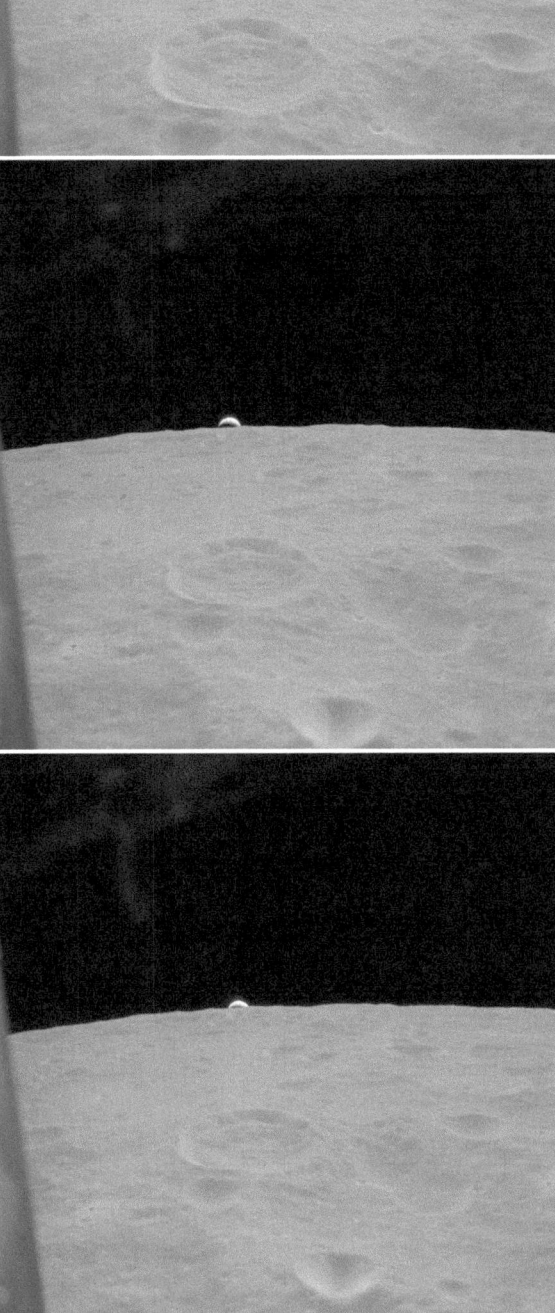

203 Earthrise over the lunar horizon for the crew of Apollo 15. The narrow bright crescent of the Earth here appears like Venus in the morning or evening sky at times of its greatest angular distance from the Sun. Apollo 15, August 1971 (Photo: Earth Science and Remote Sensing Unit, NASA Johnson Space Center, AS15-97-13267)

204 Almost a counterpart to the Earth shot of Apollo 17, which achieved cult status. About an hour after injection into the flight path towards the Moon, the crew of Apollo 16 took this photo of the fully illuminated western hemisphere. The view extends over North America and beyond the Arctic Circle. Apollo 16, April 1972. (Photo: Earth Science and Remote Sensing Unit, NASA Johnson Space Center, AS16-118-18880)

205 Two spacecraft with Earth inhabitants, on the right the natural one, home to several billion people, on the left the Apollo 16 command module with Ken Mattingly as the only inhabitant, after Charles Duke and John Young had separated in the lunar module and descended towards the lunar surface. Apollo 16, April 1972. (Photo: Earth Science and Remote Sensing Unit, NASA Johnson Space Center, AS16-113-18289)

206 During the last manned Moon flight, one of the most famous photos of the Earth was taken. It shows it fully illuminated shortly after the injection of Apollo 17 towards the Moon. The flight path allowed for the first time a view of the ice-covered Antarctica. This shot was for a long time one of the most printed images of our home planet. It gained particular cult status for the emerging environmental movement in the 70s of the last century as "Blue Marble". Apollo 17, December 7, 1972. (Photo: Earth Science and Remote Sensing Unit, NASA Johnson Space Center, AS17-148-22727)

207 Left: Harrison Schmitt of the Apollo 17 crew poses on the lunar surface in front of the American flag, above it shines the blue-white Earth. Apollo 17, December 1972. (Photo: Earth Science and Remote Sensing Unit, NASA Johnson Space Center, AS17-134-20384)

208 Right: After astronaut with flag, the Earth also adorned this shot of the lunar surface with a large Moon rock. Apollo 17, December 1972. (Photo: Earth Science and Remote Sensing Unit, NASA Johnson Space Center, AS17-137-20910)

209 Left: A view of the Apollo 17 lunar module standing in the Taurus-Littrow region of the Moon with the Earth shining above it. Apollo 17, December 1972. (Photo: Earth Science and Remote Sensing Unit, NASA Johnson SpaceCenter, AS17-134-20463)

210 One of the most beautiful Earthrises taken from lunar orbit comes from Apollo 17. The Earth showed only a narrow crescent as it appeared over the lunar horizon. Apollo 17, December 1972. (Photo: Earth Science and Remote Sensing Unit, NASA Johnson Space Center, AS17-152-23271/23274)

211 With the unmanned mission Artemis-1, launched on November 16, NASA begins a new attempt, together with international partners to land people on the Moon again with international partners. When the spaceship named Orion was already many hours on its way towards our Moon on the same day, this photo was taken from a distance of just over 90000 km. The Earth appears in it similar to the times of Apollo. November 2022. (Photo: NASA)

212 On the 13th day of its mission, Orion reached a distance from the Earth of more than 430000 km. It achieved this unusual perspective with the Moon in the foreground at a distance of 70000 km. Far behind it, the Earth can be seen. Artemis-1, 28th November 2022. (Photo: NASA)

9.   Inner Lagrange Point - In Lockstep around the Sun

The Earth moves gravitationally bound around the Sun within a year. Any mass that orbits the Sun at a greater distance takes longer for one orbit, while objects closer to the central star take less time. This simple relationship generally applies in a two-body system. However, when a third mass is involved we are dealing with a three-body problem. If we then wanted to determine the orbits of these three bodies, we would not find that an analytical solution.

The two mathematicians Joseph-Louis Lagrange and Leonhard Euler found a way out in their work on celestial mechanics in the second half of the 18th century, however, in the case that the third mass is negligibly small compared to the other two objects. They determined five points, known today as Lagrange points or also as libration points, in which, taking into account the centrifugal forces, an effective gravitational equilibrium state of the two larger masses occurs and the third small body can seemingly remain in these points. If the Sun and Earth, actually the Earth together with its Moon, form the two larger masses a small spacecraft can stay in these points and moves at the same speed as the Earth around the Sun.

Three of the total five Lagrange points, called L1 to L5, lie on the connecting line Sun - Earth. L1, the inner Lagrange point, is about 1.5 million km from the Earth towards the Sun. We find L2 1.5 million km outside the Earth's orbit in the opposite direction and L3 is located from the Earth's point of view beyond the Sun near the Earth's orbit. The two points L4 and L5 are found on the Earth's orbit and together with the Sun and Earth form an equilateral triangle.

For satellites, the first two mentioned Lagrange points are of great interest. Even though they cannot stay exactly at these points due to orbital disturbances but need continuous small corrections, these Lagrange points are considered an outstanding workplace. A spacecraft that moves around point L2 is shielded by the Earth from the Sun, but for that, it only ever sees the unlit side of the Earth. What represents an ideal perspective for astronomy satellites is therefore less suitable for Earth observations. The situation is different at point L1. There the probe is exactly between the Sun and Earth and can look at their permanently fully illuminated side. Only the apparent size of the Earth has already massively shrunk from the inner Lagrange point and reaches only an angle of about 0.5°.

In the past, there were some probes that were placed near the two Lagrange points L1 and L2. However, it was not until the summer of 2015 that the "Deep Space Climate Observatory" abbreviated DSCOVR, a satellite equipped with a camera designed for Earth observations, reached point L1. There it moves in a so-called "Lissajous orbit" around this point over the course of six months. The main task of this joint venture of NASA, the American National Oceanic and Atmospheric Administration NOAA and the US Air Force deals with the study of the Sun and its associated "space weather" Positioned 1.5 million km in front of the Earth, it can provide early warnings of approaching solar storms. The EPIC camera, short for "Enhanced Polychromatic Imaging Camera" for Earth observation, on the other hand, looks towards Earth and takes a picture of our home planet every two hours. It consists of a Cassegrain telescope with an aperture of 30 cm and a focal length of 2.90 m. The imaging CCD sensor contains 2048 x 2048 pixels. This makes it possible, even from a distance of 1.5 million kilometers, to depict the Earth in full format with surprisingly high detail accuracy in its natural color. From its data in the visible, infrared and ultraviolet spectral range, statements about the Earth's climate can be derived. The camera registers the solar radiation scattered back from the Earth like most Earth observation instruments today multispectrally in ten wavelength ranges. In July 2015, DSCOVR sent the first pictures from its unique position 1.5 million km away from Earth. At first glance, they appear like images from geostationary satellites with somewhat reduced resolution, taken at local noon. Unlike these, whose views change over the course of a day from full Earth to midnight new

Earth, the images from the inner Lagrange point always show a full Earth and the scenery changes in it in the 24-hour rhythm of the Earth's rotation, like on a rotating globe. Similar to geostationary orbits, seasonal effects can also be very well followed from the inner Lagrange point.

Depending on the position of the Earth's axis relative to the Sun, we see in the images of the EPIC camera at the times of the equinox evenly illuminated Earth hemispheres while at the winter solstice the Antarctica moves into the field of view and at the summer solstice also large areas of the Northern hemisphere are visible. Sometimes the Moon on its orbit around the Earth passes through the field of view of the camera pointed at the Earth. If it is between the satellite and Earth, we look from an unusual perspective at the illuminated back side of our companion, which is not visible from Earth. In such a case, with a suitable constellation, even a solar eclipse visible on the EPIC images can occur. It can then even be followed how the shadow of the Moon moves across the entire day side of our home planet. The appearance of the shadow cast varies depending on whether it is a total or annular eclipse. If the Moon disappears for DSCOVR behind the Earth and reappears on its other side, we see its familiar, fully illuminated front side.

213 From the distance of the inner Lagrange point not only overview shots of the Earth like from a geostationary satellite are possible. They also show a lot of detail, like for example in this shot from early September 2020. From the then raging forest fires on the west coast of North America, you can see here the smoke plumes as a yellowish-brown veil moving towards the Pacific. DSCOVR, September 10, 2020. (Photo: NASA, GSFC)

214 In July 2015, the EPIC camera on DSCOVR sent images of the Earth from a distance of almost 1.5 million km for the first time. Due to the positioning at the inner Lagrange point, the fully illuminated Earth can be observed continuously 24 hours a day. The photos received at that time show the hemispheres of all 6 continents - Europe and Africa (top) and North and South America (bottom left) as well as Asia and Australia (bottom right). DSCOVR, July 6, 2015. (Photo: NASA, GSFC)

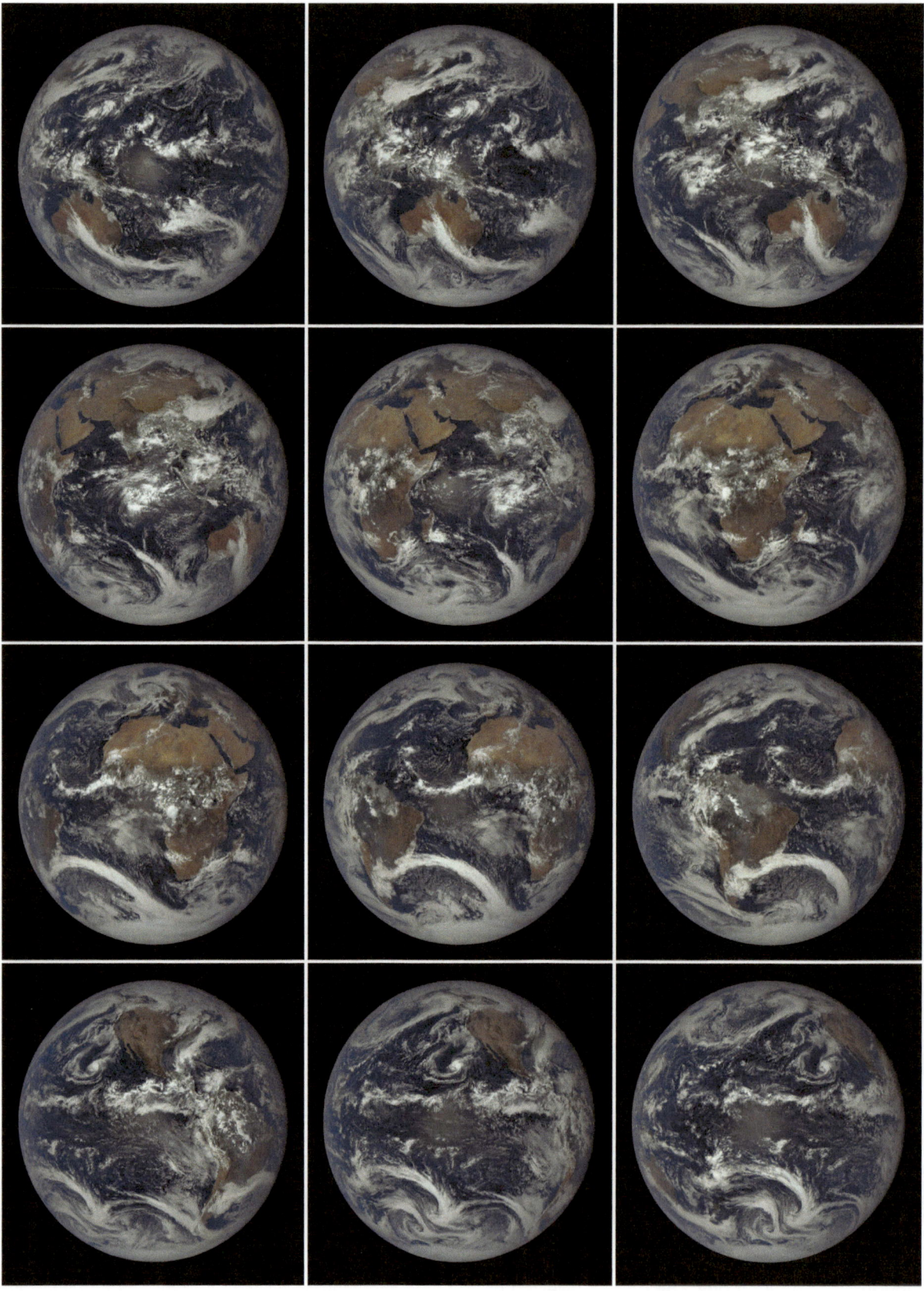

215 A rotating globe from a distance of 1.5 million km. In early October 2020, the DSCOVR satellite followed a full day on the sunlit side of the Earth in a time interval of about every 2 hours. DSCOVR, October 4, 2020. (Photo: NASA, GSFC)

216 From the inner Lagrange point, one can very clearly illustrate basic facts of celestial mechanics such as the origin of the seasons. At the beginning of autumn and spring, the equinoxes (top left and bottom left), the northern and southern hemispheres are evenly illuminated. The Earth's equator of the fully illuminated Earth therefore runs through the middle of the image. At the beginning of winter (top right), the southern hemisphere is inclined towards the Sun and can look at Antarctica while Europe is barely visible anymore. At the beginning of summer (bottom right), on the other hand, the northern hemisphere points towards the Sun. Now one can see parts of the Arctic and Europe is clearly visible. DSCOVR, 2020/2021. (Photo: NASA, GSFC)

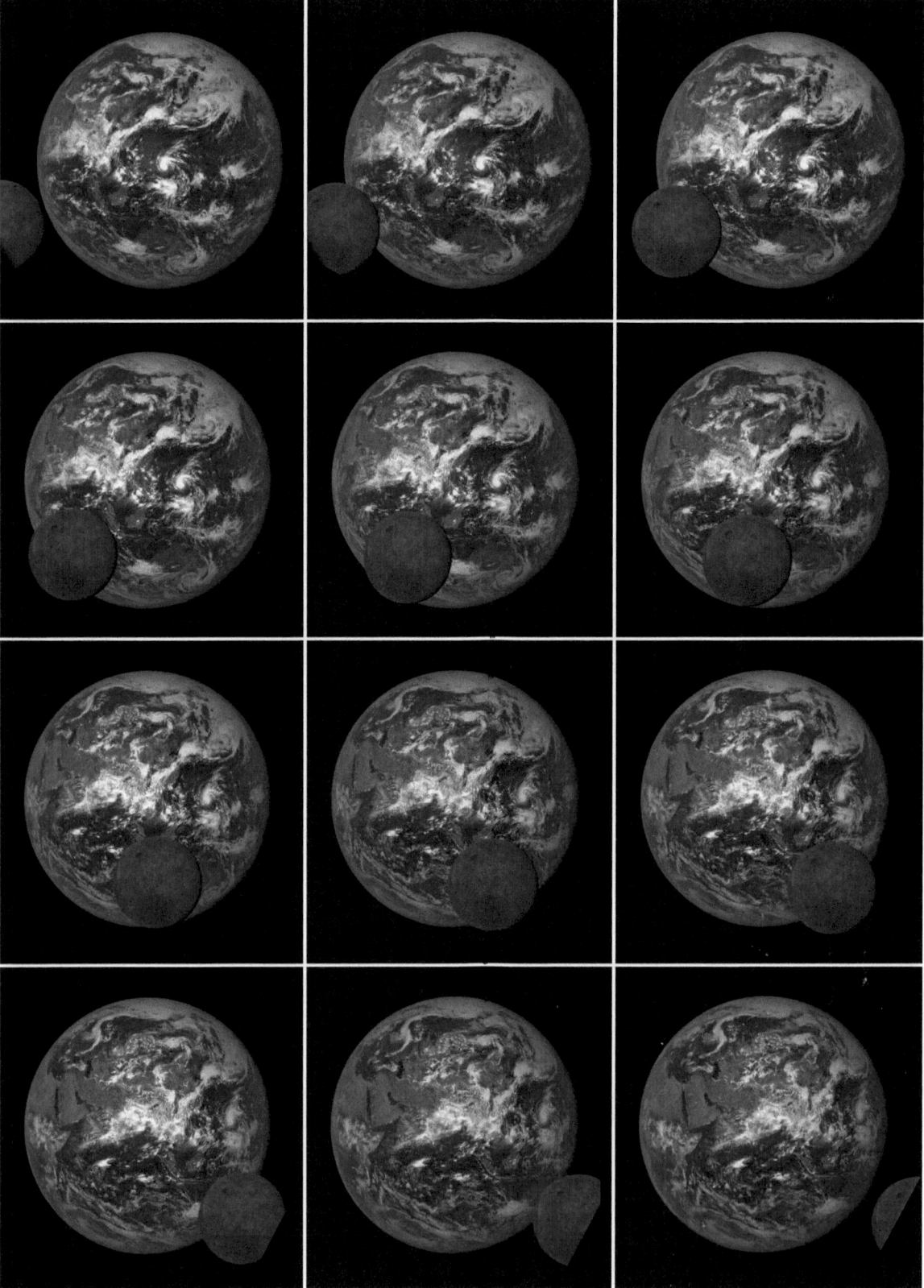

217 The Moon orbits the Earth and always shows the same side to it. When it passes through the field of view of the EPIC camera, as here in early July 2016, DSCOVR, standing far behind the Moon between Earth and Sun, always looks at its backside. The dark region in the upper half of the Moon is the Mare Moscoviense and below it, towards the left edge of the Moon, is the Ziolkowski crater with its dark crater floor. The duration of the image sequence is just over three hours. DSCOVR, July 5, 2016. (Photo: NASA, GSFC)

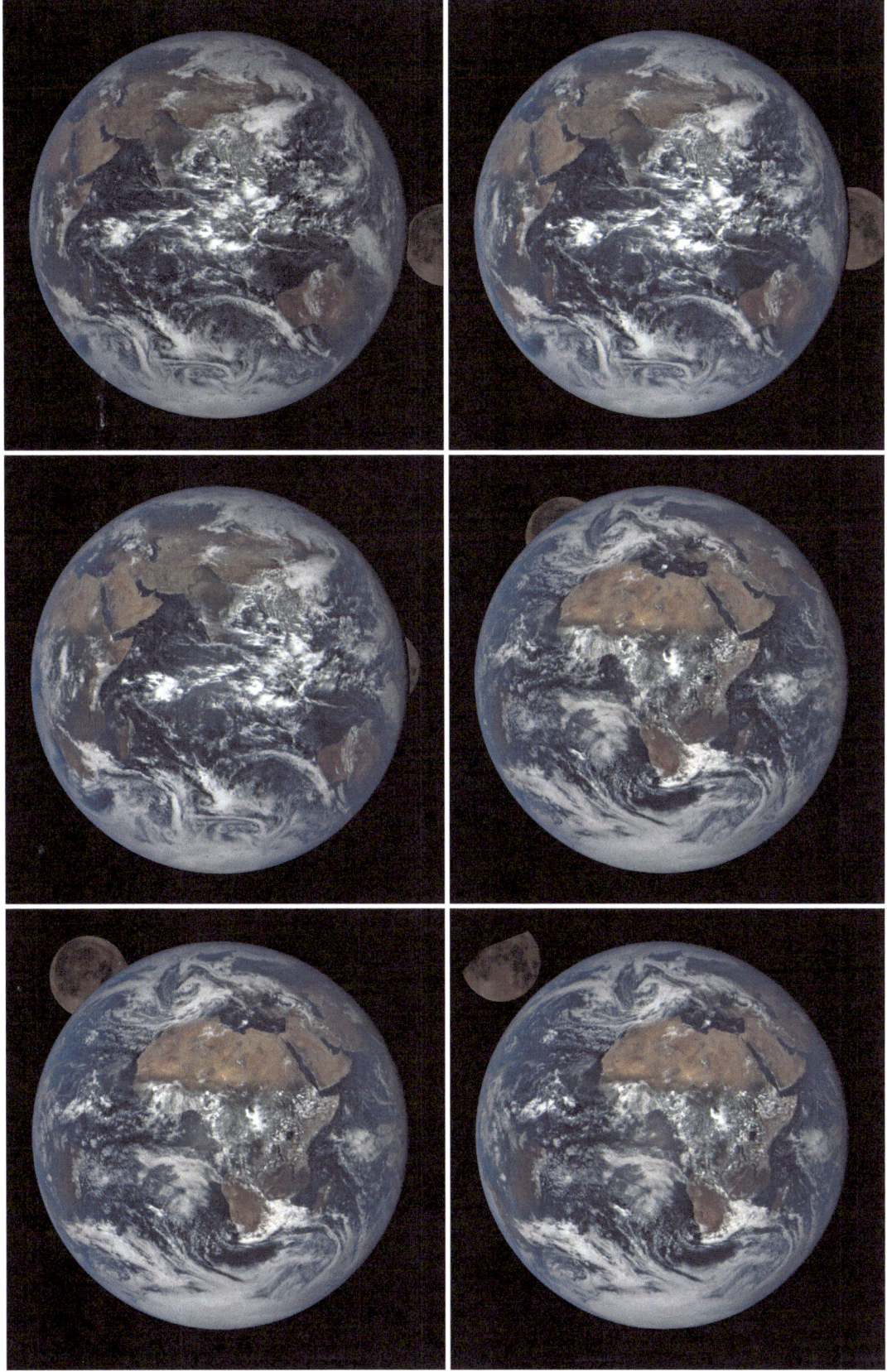

218 When the Moon passes behind the Earth in the field of view of the EPIC camera, we always see the time of the full Moon with the known structures of the Moon's front side. The sequence shown here covered a period of six hours. DSCOVR, October 2, 2020. (Photo: NASA, GSFC)

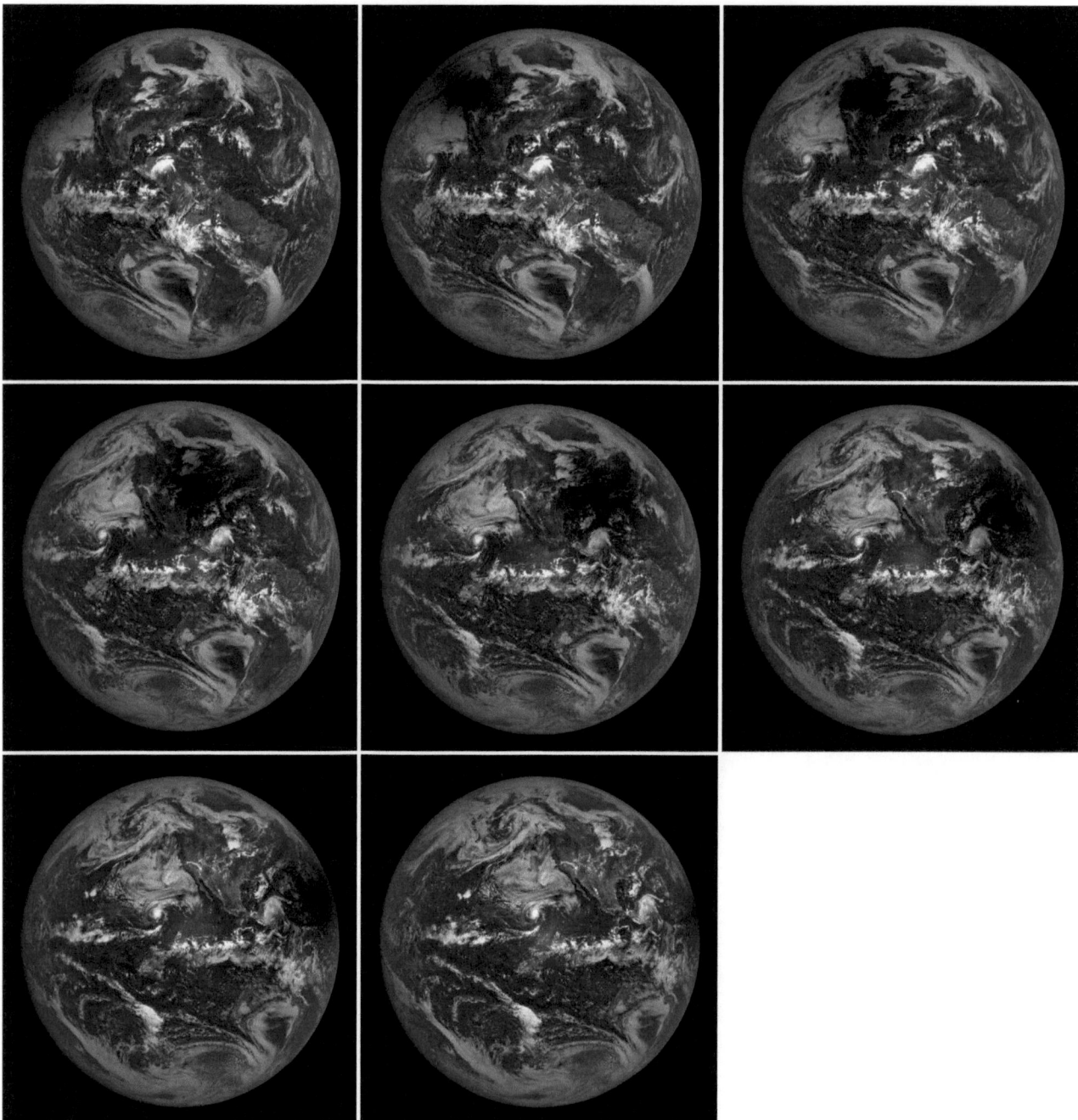

219 When the Moon is in a suitable configuration between Earth and Sun at the time new Moon, one might possibly witness a solar eclipse. From the inner Lagrange point, one then sees the Moon's shadow moving across the Earth. During a total solar eclipse the umbra is pitch black, as here in August 2017, when the zone of totality swept across the northern Pacific, the United States, and the central Atlantic. This eclipse is also depicted in the chapter on manned low Earth orbit and unmanned Moon missions. DSCOVR, August 21, 2017. (Photo: NASA, GSFC)

220 A rather rare event is a total solar eclipse directly over the South Pole. Such an event could actually be observed in December 2021 over Antarctica, when the umbra of the Moon crossed its western part. DSCOVR, December 4, 2021. (Photo: NASA, GSFC)

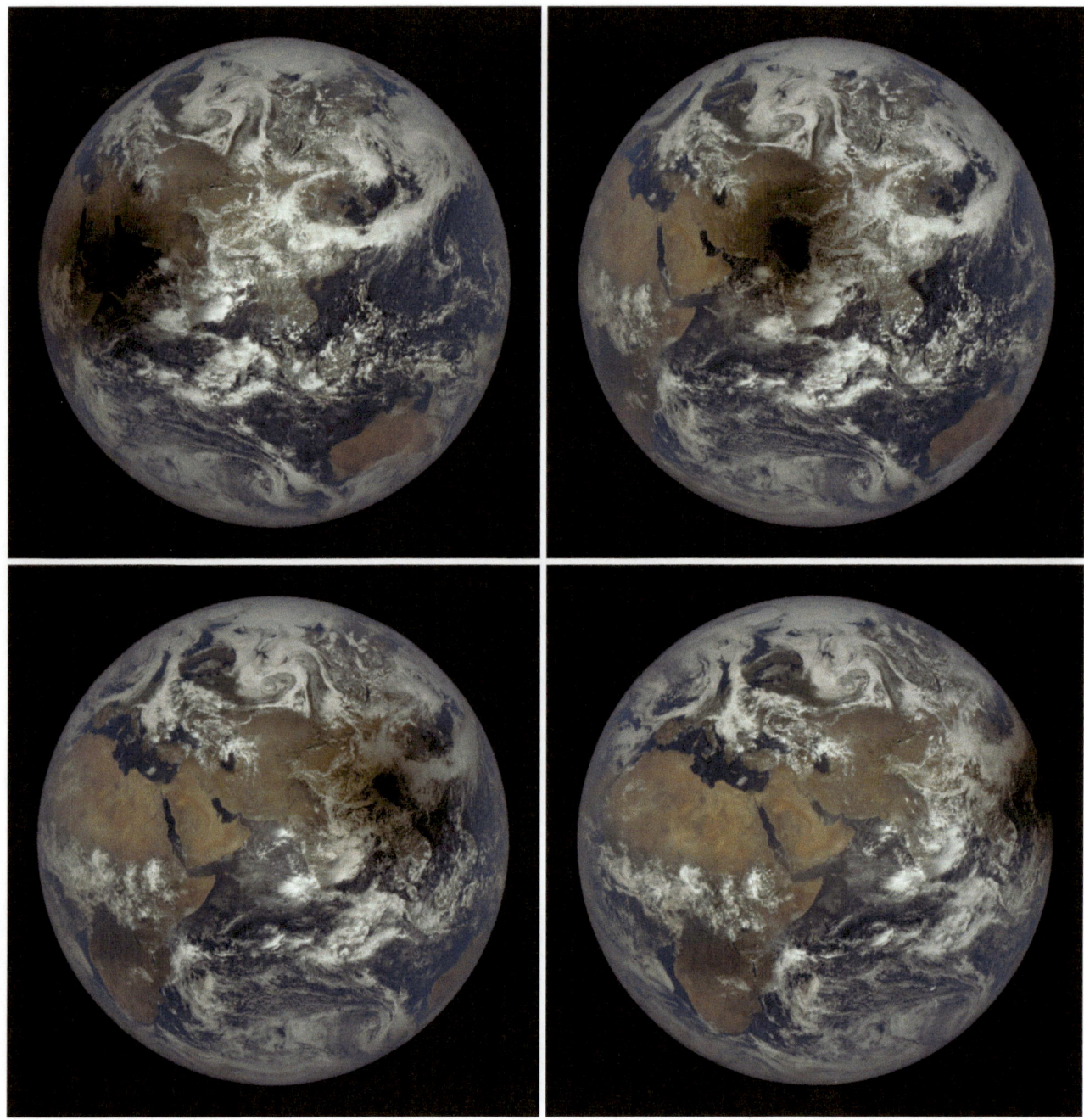

221 During an annular solar eclipse, the Sun does not completely disappear behind the Moon. This is evident in the Moon's shadow of the annular eclipse from June 2020 with a coverage degree of over 99%, which still appears yellow-brown and not pitch black. This solar eclipse was mainly visible in Africa and Asia. DSCOVR, June 21, 2020. (Photo: NASA, GSFC)

222 Exactly 50 years after Apollo 17 took its photo of the Earth, which gained cult status, the EPIC camera on DSCOVR also captured this scene with Africa prominently on the left edge of the image and further down in the south the Antarctica. The slightly different position of the continents is due to the different shooting perspective. Apollo 17 was about 29000 km above the Earth while the EPIC shot was taken from a distance of 1.5 million km. DSCOVR, December 7, 2022. (Photo: NASA, GSFC)

10. Interplanetary Travels - Visits to our Cosmic Neighborhood

Beyond the Moon begins interplanetary space. The further one moves out into it, the smaller it appears from there the Earth and eventually resembles the other large planets of the solar system, as they can be seen from Earth either through small telescopes as discs, even with details on the surface and in the atmosphere, or with the naked eye as bright objects in the sky. Already early in the space age, it was possible to send spacecraft - albeit uncontrolled - into the interplanetary area. December 1962 finally marked the beginning of the exploration of our solar system when NASA managed to fly a probe controlled near the planet Venus for the first time. Since then, just over six decades have passed, during which numerous space probes, many successful, some less successful, visited our large and small relatives in the solar system up to its edge. Incidentally, sometimes there was time and opportunities to also take a look back at Earth. The choice of motif was simple. Either you saw the Earth alone in different illuminations conditions, or you photographed the Earth-Moon system in various constellations. In the age of digital image processing, one has the freedom to design the images in such a way that certain aspects stand out better to the viewer. You can adjust colors, highlight the Moon with its very low albedo a little more clearly or enhance the brightness of the Earth when it is barely visible in the image from a great distance. But even in such cases, we always get a image that corresponds to the natural view.

What one sees in it, of course, depends on the target of the probe. The further one ventures into the outer solar system, the smaller our home planet appears.

Travel Routes

If you want to visit another planet from Earth, it is necessary to first bring a space probe into an Earth orbit, from there to then embark on the flight towards the planet. For this, one always has to achieve at least the terrestrial escape velocity of 11.2 km/s which allows the probe to overcome Earth's gravity and advance into interplanetary space. After that, the probe moves on an elliptical orbit in the gravitational field of the Sun. In the simplest case, this ellipse will just reach the orbit of the visited celestial body around the Sun. Since one knows its orbital motion, as well as that of the space probe, one can choose their departure from Earth in such a way that the probe and celestial body briefly meet, after which the probe follows its path dictated by the gravitational force of the Sun, which brings it back towards Earth. If the spacecraft is to remain in the vicinity of the visited planet and perhaps even swing into an orbit there, one must align the velocities of the planet and the space probe. The Earth orbits the Sun at an average speed of 29.8 km/s. All outer planets move slower, starting with Mars at 24.1 km/s up to Neptune at 5.4 km/s. Just as follows from Kepler's third law, the inner planets Mercury and Venus must orbit the Sun faster. Venus reaches 35 km/s and Mercury even 47.8 km/s. In reality, the planetary speeds vary within a certain range, depending on how eccentric the respective orbit is. To now fly from Earth to one of the outer celestial bodies, the probe initially needs a higher orbital speed than that of the Earth, to move further outward in the gravitational field of the Sun; it must be accelerated. On its further path,

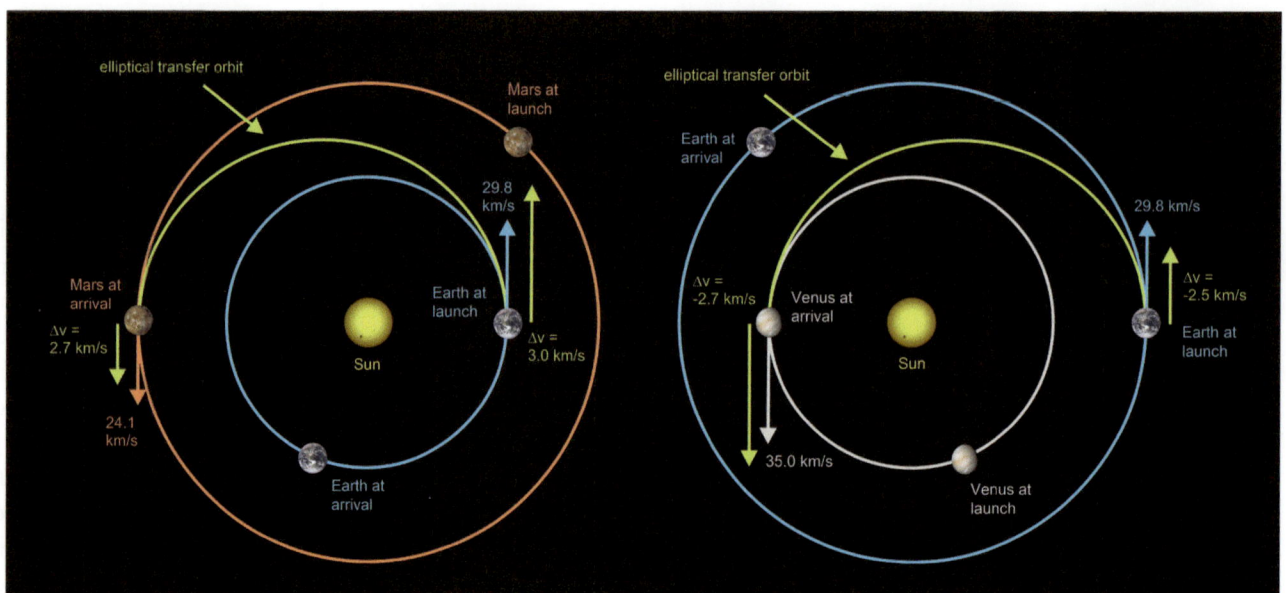

223 Schematic image of the energetically favorable Hohmann orbits to Mars as a representative of an outer planet (left) and to Venus into the inner solar system (right). For simplicity's sake, the orbits of the planets are assumed to be circular. The indicated speeds are relative speeds to the Sun. For a flight to Mars, the probe needs a speed to overcome the attraction of the Earth and leave the Earth's orbit. It also has to be higher than the Earth's orbital speed, so that it hits the orbit of Mars at the farthest point from Earth. With the correct choice of the starting time, Mars is also at this location. However, the probe has now lost speed. It is only 21.4 km/s and would be left behind by the passing Mars while the probe would move back towards Earth. An additional thrust, which results in a speed change of 2.7 km/s, keeps the probe in step with Mars and allows further maneuvers to swing into an orbit. The flight from Earth to Mars outlined here would take about 260 days. In the other direction, into the inner solar system, it is similar, only with the opposite sign. A probe to Venus must again be accelerated to a speed to leave the Earth's orbit and the Earth's gravitational field. However, its relative speed to the Sun must now be smaller than the Earth's orbital speed. Then the probe "falls" towards the Sun to hit the orbit of Venus at its farthest point from Earth. Here too, by choosing the correct starting time, Venus can also pass by at this point. Now, however, the probe has gained speed and would overtake Venus at 37.7 km/s to then move back towards Earth. A braking maneuver must reduce the speed of the probe as a prerequisite for entering an orbit around Venus. A flight to Venus as described here would only take about 150 days. The direct reaching of the orbit of another planet described here is the energetically most favorable variant to approach it from Earth. This was investigated and published by the German civil engineer Walter Hohmann in the 1920s. In his honor, such transition orbits are called "Hohmann orbits" Other elliptical orbits are also possible, but with higher effort.

it will now slow down and reach its lowest speed at the point furthest from the Sun, where it meets the planet and would then "turn around" at its lowest speed. Arriving at Mars, it now only moves at 21.4 km/s. An additional thrust of about 2.7 km/s is required, so that the probe continues to stay near Mars. The arrival and stay at more distant planets would require significantly higher accelerations.

Flights to the inner planets behave in principle the same, only with reversed sign. To get to the inner solar system, the speed of the spacecraft when leaving Earth must be reduced For a flight to Venus, it would require a speed of 2.5 km/s, and even 7.5 km/s for Mercury. It then "falls" on an elliptical path towards the Sun and reaches 37.7 km/s at its turning point at the distance of Venus

and already 56.4 km/s at Mercury. After that, it would move back outwards towards Earth. Now, another deceleration is necessary, only then do the inner planet and probe follow their course around the Sun. Additionally, the travel time plays a role. Venus and Mars can be reached on the described paths in significantly less than a year. It already takes almost three years to get to Jupiter and you only get close to Neptune at the edge of the solar system after 30 years.

In space travel, it is important to move as effectively as possible. Any change in speed means the explosive combustion of fuel with conventional rocket engines. The greater the change, the more fuel is needed; the more fuel, the more mass has to be launched from Earth, which in turn requires

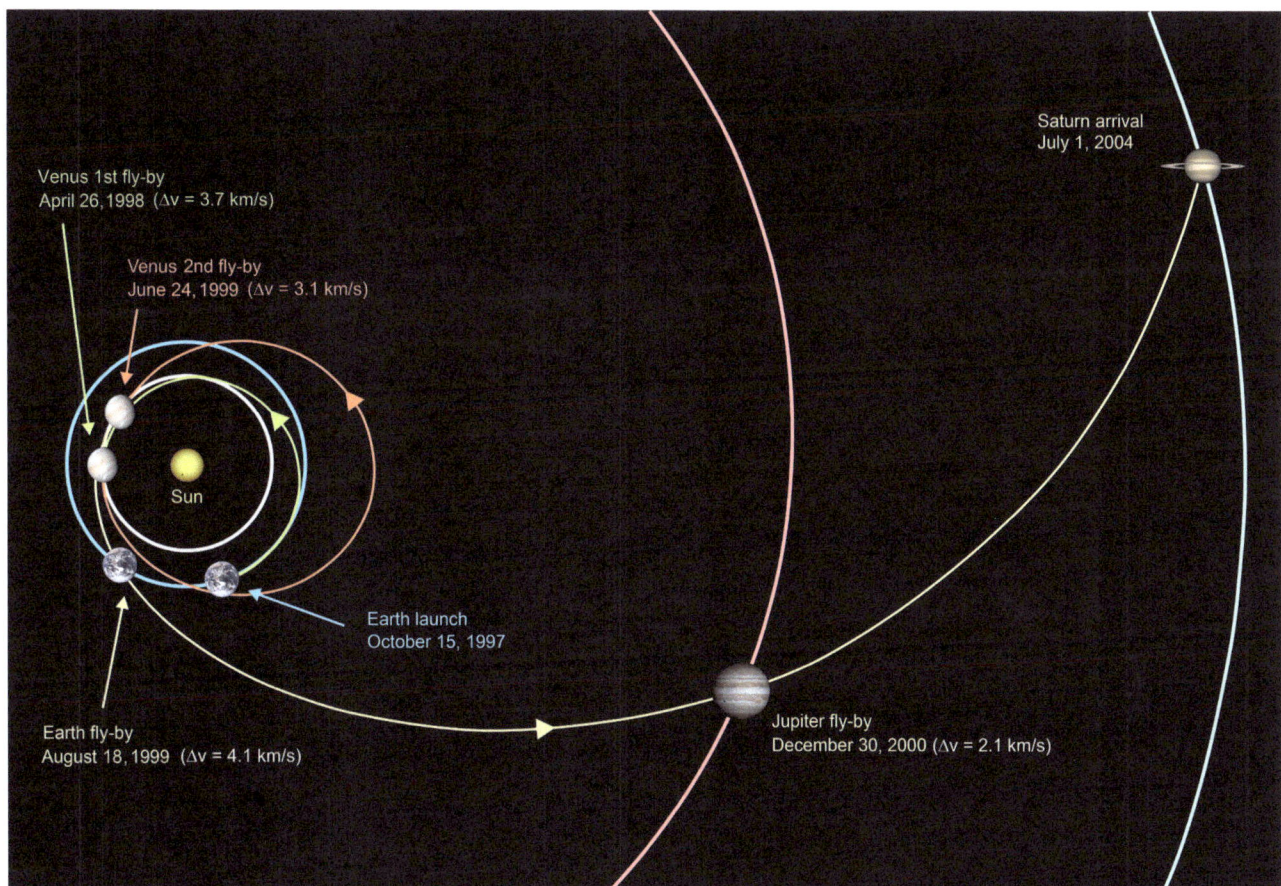

224 A flight to an inner or outer planet using gravity maneuvers (gravity assist, swing-by, or fly-by) proceeds quite differently, as illustrated here by the example of the nearly seven-year-long flight of the Saturn probe Cassini-Huygens to the ringed planet. On a Hohmann trajectory, the flight to Saturn would have taken only a little longer than six years, but even the most powerful carrier rocket of NASA could not accelerate the heaviest interplanetary probe to the speed required for a direct flight. Therefore, four flybys of planets were planned: two at Venus, one at Earth and one at Jupiter. The flight phases are color-coded in the graphic: green from the start to the first Venus flyby, amber between the first and second Venus fly-by, and yellow from the second Venus fly-by until arrival at Saturn. The speed change relative to the Sun achieved at each gravity assist is also indicated.

more powerful launch rockets - or the probe is reduced. With today's rocket engines, only direct flights to Venus and Mars are possible, whereas for all other planets the necessary high amounts of fuel for twice accelerating or decelerating prevent the launch of an interplanetary, well-equipped probe. Since we have already visited a large part of the solar system today - a fly-by of Neptune took place in the summer of 1989; Pluto and a member of the even further outlying Kuiper Belt were passed in 2015 and 2019 respectively – there must be another way for interplanetary travel.

The foundations for this were laid out by Michael Minovitch in 1961 when he was a student working for NASA. His method is known as "gravity assist" or "gravity maneuver". Sometimes it is also referred to as "swing-by" or "fly-by" It allows the speed of a space probe to be changed without fuel consumption, only by cleverly chosen close fly-bys of planets or even large Moons. When a spacecraft on its ellipse around the Sun has approached a planet closely enough, its gravitational pull predominates, the space probe is deflected in its path and accelerated. After a closest approach, it moves away again in the gravitational field of the planet, slows down, but has changed its direction. Likewise, its speed relative to the Sun no longer corresponds to that which it had before the fly-by. The planet orbiting the Sun and the space probe have exchanged momentum and energy during the encounter. For the much more massive celestial body, the influence is negligibly small, the light probe, on the other hand, has experienced a significant speed and direction change. In fact, during the fly-by, the sum of the kinetic energies and the momenta of probe and planet remain the same. It now depends on which side the space probe approaches the planet. Depending on the approach direction, you can achieve either an acceleration or a deceleration. By cleverly stringing together gravity assists at different planets, a probe can be given multiple speed changes, which ultimately add up to such high speeds that the probe can be transported to the outer solar system in a relatively short time. They can also become so large that even the gravitational field of the Sun is no longer sufficient to keep the space probe in our solar system. After a long enough time, such spacecraft leave the interplanetary space and reach the interstellar space of our Milky Way. But also Mercury as the planet closest to the Sun becomes accessible through gravity assists. They enable a probe to be slowed down so much that it can enter an orbit around Mercury upon arrival. Interplanetary spacecraft can also gain energy at Earth. In such a case, they cross the Earth's orbit on their path around the Sun at the exact moment when our home planet passes by. Often, the onboard cameras are then tested and images are obtained that resemble images from Moon distance, from geostationary orbit or even from low Earth orbits. Not all photos of interplanetary probes show the Earth from very large distances. In the previous chapters, we proceeded largely chronologically. Each stage away from Earth began with the first painstaking successes and ended with today's outstanding results. Listing the interplanetary flights with Earth view using this approach would lead to a considerable back-and-forth in the solar system. The journeys to the inner and outer planets were never ordered in time. It is better to consider each destination separately now. There are members of the solar system that have only been visited once or twice from Earth, others, on the other hand, have been receiving visits from Earth since the beginning of interplanetary space travel, with increasingly complex mission objectives.

The next planetary neighbors

Mercury, Venus, and Mars, the other members of the inner solar system besides Earth, are considered "terrestrial" or "Earth-like" due to their solid, rocky surface. Venus and Mars are the closest to Earth. In

the best case, only 42 million km extend between Earth and Venus or 55 million km between Earth and Mars. Both planets can be flown to directly. Mercury, the innermost planet, can come as close as 80 million km to Earth at minimal distance. Despite this actually small distance, it is practically unreachable on a path as described by Walter Hohmann with a swing into an orbit there. Only with the help of swing-by maneuvers are the necessary speed changes achieved. This, and the fact that Mercury as a planet without an atmosphere shows less Earth similarity, led to the fact that so far only three probes have been sent to it. The beginning was made at the end of 1973 by Mariner 10. NASA steered Mariner 10 past Venus to perform the first gravity maneuver in the history of interplanetary space travel, which maneuvered the probe into an orbit in which it passed close to Mercury three times and completed its scientific program. Shortly after the launch, when the Earth had already been left, the Earth-Moon system was recorded for the first time from a distance of 2.6 million km beyond the Moon's orbit - and this even in high resolution and in color. To illustrate the correct size and color differences, both celestial bodies were arranged side by side in the resulting image.

It took more than 30 years until in August 2004 Messenger, another NASA vehicle, set course for Mercury. This probe needed significantly more gravity assists, namely six. It passed once by Earth, twice by Venus, and three times by Mercury before its speed was reduced enough that a final ignition of the engine was sufficient to swing into an elliptical Mercury orbit in March 2011. After launch, Messenger followed a path that brought the probe close to Earth for a swing-by a year later. Already in May of the same year, Messenger's high-resolution dual camera system had been activated and Earth and Moon had been documented from 30 million kilometers. The same happened a few months later when the spacecraft passed Earth at a distance of only about 2500 km. Numerous photos show the close fly-by in August 2005 over the course of a day. A video sequence created from this gives an impression of how a passenger coming from interplanetary space and disappearing back there would have experienced the fly-by. Even before entering a Mercury orbit, Messenger looked towards Earth several times. Between the Sun and Mercury's orbit, the search for theoretically postulated, but so far undetected minor planets, the Vulcanoids, was on Messenger's measurement program. In May 2010, the Earth-Moon system also came into view from a distance of 183 million km. In the wide-angle shot, both celestial bodies stand next to each other as bright disks against the star-studded background, showing parts of the constellations Libra and Scorpio, due to overexposure. Six months later, in November 2010, NASA attempted a "family portrait" of all the planets of the solar system, as had already been created by Voyager 1 in 1990, but then from beyond the orbit of Pluto. In individual shots, all planets along the ecliptic were photographed in a panoramic view. Not only did the Earth show its Moon; several Galilean Moons of Jupiter can also be recognized. Later, on July 19, 2013, when Messenger was already in orbit around Mercury, NASA even organized a terrestrial photo campaign together with the Saturn probe Cassini. From different directions and distances, Messenger 100 million km and Cassini more than ten times as far, the Earth was photographed. The date was announced, so that every interested person on Earth could look up at the sky and be sure to be in the field of view of one of the two probes. The resulting photo strongly resembled that of May 2010, but Earth and Moon appeared strongly overexposed.

Lunar eclipses are an interesting celestial spectacle. This applies not only from the Earth's perspective, but also from the perspective of another planet. Messenger observed such an event on October 8, 2014 from Mercury's orbit. In the sequence of images, you can see how the Moon slowly disappears into the Earth's shadow within an hour. On Earth, observers in North America and the Pacific region were able to follow the phenomenon. Since October 2018, a joint venture initiated by the European Space Agency ESA together with its Japanese equivalent JAXA under the name "BepiColombo" is on its way to Mercury. It is expected to enter an orbit around Mercury at the end of 2025 after six gravity maneuvers at Earth, Venus, and Mercury. In April 2020, a close flyby of Earth already took place and could, similar to Messenger, be documented photographically in black and white. Venus is a much more thankless object for camera lenses than Mercury. Its dense atmosphere prevents a view of its surface. Structures can only be discerned in the upper atmospheric layers and

there only in the ultraviolet spectral range. If you want to unravel the secrets of Venus, you have to resort to special observation methods such as radar technology, which then reveals the surface of our nearest neighbor. Although Venus is one of the most visited planets - in the 70s and 80s, the USSR space program achieved twelve successful visits in a row - none of these probes carried instruments that also allowed occasional views of the Earth. If cameras were on board, they were used to inspect the surroundings of the landing site after successful touchdown on the Venus surface. Even NASA missions or the long-term European project "Venus Express" were no exception, except for the very coarse images of Earth by a spectrometer on Venus Express. Only Akatsuki, a Venus orbiter operated by Japan, attempted high-resolution Earth images. After its launch on May 20, 2010, ultraviolet- and infrared-sensitive cameras photographed the Earth from a distance of 250,000 km, but not at visible wavelengths. By the end of the year, Akatsuki reached Venus; however, the attempt to enter an orbit around the planet failed and was postponed to a later date at the next meeting of the probe and Venus. This actually succeeded as planned at the end of 2015.

Unlike Venus, Mars allows an unobstructed view of its surface. Its thin atmosphere, only occasionally disturbed by dust storms, does not pose an insurmountable obstacle for telescopes and cameras. Mars is the only planet where telescopes can make out surface details. The seasonal changes observed led to early speculations about life on our neighbor. When Mariner 4 first got a look at Mars in 1965, there was great anticipation about what it would show from close range – and in a way, disappointment when Moon-like crater landscapes were presented. Today, after more than six decades of space travel to Mars, the situation is seen more differentiated. Mars may have had a more Earth-like character in its early days; its further development then turned it into the cold dry world we see today. In pictures that probes send us from Mars orbit or from its surface, we see desert-like scenes that sometimes seem strangely familiar to us. It is probably due to this apparent Earth-likeness that makes the views from Mars back to Earth particularly appealing: We look from a world that failed to become a habitat for developed life to our planet, whose evolution has maintained the suitable environment for billions of years. Although we have made more than twenty successful visits to Mars since 1965 using space probes, the first image of Earth from a platform moving towards Mars was not achieved until 1998. It came from the Japanese probe Nozomi ("Hope") which was launched in July of the same year but was unfortunately lost due to defects. From a distance of 160,000 km, the photo showed the Earth-Moon system. A few years later, Mars Odyssey had more success. Upon leaving Earth after its launch in April 2001, the onboard instruments were tested, including a camera capable of taking images in the infrared spectrum. It shows the Earth as a bright disc and the Moon as a significantly smaller companion - and nothing but empty space in between. In the infrared image, the full Earth shines due to its thermal radiation, while the corresponding image at visible wavelengths only shows a narrow Earth crescent. From a similar position, but already 8 million km away from Earth, the European Mars Express looked back two years later in July 2003 to see the two celestial bodies. The different scale compared to the image from Mars Odyssey is only due to the different imaging optics. Not only the recording distance determines the apparent size of the Earth, but also the focal length and the sensor of the camera contribute significantly to it.

In May of the same year, the Mars Global Surveyor, launched by NASA in 1996 and orbiting Mars since September 1997, experienced a constellation where Mars, Earth, and Jupiter were in a line. The distance from Mars to Earth was 139 million km and to Jupiter, further behind, 944 million km. In two images, Earth and Moon as well as Jupiter together with its bright Galilean moons were photographed. The sight of the gas giant looked like it was viewed from Earth through a small telescope while the half-illuminated Earth was detailed for the first time from a neighboring planet. It had taken almost four decades since the first image of Earth from lunar distance in 1966 by Lunar Orbiter 1 had been achieved. Earth and Moon were in almost identical positions on October 3, 2007, when the Mars Reconnaissance Orbiter, equipped with a very high-resolution camera system, sent a photo to Earth that strongly resembled the one from May 2003 of the Mars Global Surveyor but was of significantly higher quality. It shows with what improved camera the Mars Reconnaissance

was equipped when the orbiter was sent to Mars in 2005.

So far, nine autonomous laboratories have successfully landed on Mars. Six of them were even mobile and could be controlled from Earth to move across its surface. Occasionally, the mounted cameras also dared to look upwards, especially at times of twilight. In such an attempt in early March 2004, the Mars Exploration Rover named "Spirit", just 63 days after it had landed, captured the Earth as a brightly shining object above the Martian horizon, an hour before sunrise. A few months after the first Earth image from the orbit of another planet was achieved, there was now also a photo of our home from the surface of another planetary neighbor. Later, images were added that show the Earth together with Venus in the twilight of Mars. However, little notice was taken of this, as the rover got stuck in the sand at the same time and had to end its mission. Spirit's Earth portrait was repeated with the much more modern mobile robot "Curiosity" who landed in Gale Crater in August 2012, it had reached an area named "Moonlight Valley" by early 2014. There, on January 31, 80 minutes after sunset in the west, it saw the Earth as an evening star above the horizon in the twilight. In full resolution, the Moon also clearly appears as our satellite. More than six years later, the Earth once again came into the camera's field of view on "Curiosity", this time even accompanied by the much lower Venus, which can be seen just above a rock outcrop.

Asteroids and Comets

Even though one might assume that interplanetary space is empty, there are still a multitude of small celestial bodies - meteoroids, asteroids, and comets - bound by gravity, they trace their orbits around the Sun. Today, they form a highly interesting astronomical research topic, as their existence dates back to the beginnings of our solar system, and we can therefore learn something about its formation. Our space travel capabilities have now advanced considerably; they allow us to approach such objects. Their very low gravitational field or unusual shape are no longer an obstacle to entering an orbit or even landing on them after the probe has arrived. Since the visit of Halley's comet to the inner solar system in 1986, which was documented by an international fleet of probes, space missions to small bodies have become part of the repertoire of space-faring nations. The approaches usually use gravity maneuvers, often also by close fly-by of the Earth. The first small body probe, NEAR Shoemaker ("Near Earth Asteroid Rendezvous") and posthumously named after the geologist Eugene Shoemaker who died in a car accident - set off for the near-Earth asteroid Eros ten years after Halley's visit. In January 1998, the probe came as close as 540 km to Earth and was thereby put on the correct course towards Eros, where it later entered an orbit around Eros 255 million km away from Earth and studied the asteroid on site for a year. The Earth flyby in 1998 took NEAR Shoemaker over Africa and allowed a rare view of the entire Antarctica when leaving Earth. The last image of the rendezvous with our home planet was taken from a distance of 400,000 km and shows the now familiar constellation of half Earth and half Moon. Hayabusa ("Peregrine Falcon") a project of the Japanese space agency, also needed gravity assistance from Earth. Launched in May 2003, Hayabusa was to fly to the small, Earth-crossing asteroid Itokawa, orbit it, even collect material from its surface and bring it back to Earth. Despite damage to the spacecraft and therefore limited operational capabilities, this ambitious goal was actually achieved when a small return capsule, released an hour before the end of the mission, safely landed in Australia in June 2010 while Hayabusa plunged towards Earth and burned up in its atmosphere. During the gravity maneuver in May 2004, Hayabusa managed to take Earth images from different perspectives. Success spurs on. Therefore, Japan sent the successor probe Hayabusa-2 to the also Earth-crossing asteroid Ryugu in December 2014. A year later, a photographically documented swing-by maneuver at Earth took place. After successfully completing its tasks at Ryugu, Hayabusa-2 has since been on its return flight to Earth, where the return capsule with the samples of the asteroid landed successfully in Australia almost exactly six years after its launch. The mother probe is later expected to investigate two more asteroids in 2027 and 2031.

After Europe's great success in 1986 with the Giotto probe to Halley's comet, ESA had chosen another comet as the next small member of the solar system, the short-period comet Churyumov-Gerasimenko. The launch of the Rosetta spacecraft took place in March 2004. Four gravity maneuvers were needed alone - three at Earth, one at Mars - to let Rosetta meet the comet in August 2014 and, after deploying a lander, accompany it for a year on its further flight into the inner solar system. The Rosetta mission ended on September 30, 2016 when the probe was crashed onto the comet. Rosetta's advanced camera system provided spectacular images during all three Earth passages in 2005, 2007, and 2009. The image taken on the night side is impressive, with the southern hemisphere of the Earth only visible as a narrow crescent while the lights of Europe, the eastern Mediterranean, and the Middle East appear in the north.

The NASA EPOXI mission's images of Earth are also spectacular, but in a different sense. EPOXI was actually the "recycling" of the Deep Impact probe, which had successfully investigated Comet Temple 1. Since it continued to function, it was given two new tasks: visiting another comet and searching for extrasolar planets by observing stars. In May 2008, it was possible to test on Earth how an exoplanet might appear from a great distance. From a distance of 50 million km, EPOXI exposed a sequence of Earth images. Even though the Earth appeared much larger than an exoplanet can ever be depicted, something was learned about possible signatures of a life-friendly environment in the registered spectrum. The constellation on the day of the recordings even made it possible to follow the Moon's transit in front of the Earth's disk. The OSIRIS-REx probe also received the necessary velocity change at Earth in September 2017 to allow it to reach the near-Earth asteroid Bennu in December 2018, from which it is supposed to bring soil samples back to Earth. The camera systems of OSIRIS-REx followed this moment of greatest Earth proximity and provided a view back to the Earth and Moon system from a distance of more than 1 million km. Later, when OSIRIS-REx had already reached the approximately 500 m large asteroid Bennu by the end of 2018, another image was taken with the navigation camera. It shows Earth and Moon 114 million km away as faint points of light while the asteroid Bennu is bathed in dazzling light only 47 km away. In October 2020, the sample collection on Bennu was finally successful. After a further five-month stay at Bennu, the OSIRIS-REx probe left the asteroid and set off on its return journey, where it is expected back in September 2023. The most recent image of Earth from a journey to asteroids in the solar system comes from the NASA probe Lucy, named after the partial skeleton of a hominid of the species Australopithecus afarensis, which is about 3.2 million years old. Lucy's task is to investigate six Trojan asteroids that follow Jupiter at a distance of 60°. Two Earth fly-bys are necessary for this. The first took place on October 16, 2022, exactly one year after Lucy's launch, and was documented by Lucy's onboard camera. The second is then expected to follow on December 13, 2024.

Gas and Ice Giants

Relatively early on, ventures were made into the areas beyond Mars and the asteroid belt, where the gas giants Jupiter and Saturn and the ice giants Uranus and Neptune orbit. Pioneer 10 began its journey to Jupiter as early as March 1972, followed just a year later by Pioneer 11 with planned fly-bys of Jupiter and Saturn. At their destinations, the probes were accelerated so strongly by gravity assists that they could no longer be held in the solar system by the Sun's gravity on their further journey. They are gradually moving towards interstellar space. Both probes were not intended to devote themselves to Earth when the opportunity arose. Unlike their subsequent twin missions Voyager 1 and Voyager 2. These were to take advantage of a favorable position of the outer planets to explore as wide an area of the outer solar system as possible with relatively little effort, only through gravity maneuvers in a "Grand Tour". The instrumentation of Voyager 1 and Voyager 2 now allowed high-resolution images. An impression of their performance was obtained shortly after the launch of Voyager 1 in September 1977, when Earth and Moon were photographed together for the first time from 11.67 million km. Together with the probe Voyager 2, which was launched two weeks earlier, Voyager 1 provided detailed insights into the worlds of the gas and ice giants in the following years. Voyager 1 flew close to Jupiter and Saturn and their Moons, Voyager 2 reached the same targets a little later and even made it to Uranus and Neptune. As in the case of the Pioneer probes, the speeds achieved during the fly-bys were sufficient to leave the Sun's gravitational field and advance into interstellar space. The distance to Voyager 1 has now reached more than about 24 billion km. Thus, the Probe the farthest object created by human hands. Not quite as far, but still a considerable 6.4 billion km away, was Voyager 1 on February 14, 1990. In a look back from above the ecliptic, Voyager 1 attempted a mosaic of the solar system to find all the planets in it. Mercury was too close to the Sun. Its scattered light also prevented the imaging of Mars. The other six planets, including Earth, could be seen, however. Due to the small apparent distance to the Sun, our home planet is also embedded in scattered light and appears as a tiny, no longer resolved, bluish shimmering point. Since this image, the expression "pale blue dot" has been coined for the Earth when it presents itself to us from a great distance, without recognizable details. Both the Pioneer and Voyager missions allowed only brief snapshots of the outer planets during their fly-bys. A detailed study of the giant planets and their moons required probes that could be steered into an orbit. In the case of Jupiter, this was initially Galileo, a NASA mission with participation from the German Aerospace Center (DLR), which ended in 2003. The launch of Galileo took place in 1989 from the cargo bay of a Space Shuttle. Three gravity assists, two of them at Earth in 1990 and 1992, led Galileo towards Jupiter, where the probe arrived in 1995 and dedicated itself to the exploration of the Jupiter system until mid-2003. The two Earth fly-bys took place in December and were followed photographically. From the individual images, film sequences could be generated, which documented the Earth's rotation or the course of the Moon around the Earth. Carl Sagan, one of the leading figures in the popularization of astronomy in the last century, also derived information from the data obtained during Galileo's first approach to Earth that contained indications of life on Earth. This method is now considered pioneering work for future attempts to find life-friendly exoplanets using remote sensing. Although Galileo was very successful, many questions about Jupiter remained unanswered. These are to be clarified by NASA's Juno probe, which was launched in August 2011.

Shortly after its launch, the wide-angle camera was tested, which provides the images of its atmosphere near Jupiter. The only target in this case was the Earth-Moon system from more than 9 million km. Two years later, in October 2013, Juno approached Earth for the only gravity assist maneuver to then continue flying towards the largest planet. The period until reaching the minimum distance of 560 km off the south coast of Africa was captured with the wide-angle camera, after which Juno quickly moved away. Juno's arrival at Jupiter took place in the summer of 2016. Since then, the probe has been investigating the largest gas giant in the solar system from its orbit. The crash into the Jupiter atmosphere was initially planned for mid-2021; however, NASA has extended the Juno mission by another 42 Jupiter orbits until September 2025.

The second gas giant, Saturn, has also already been the target of a complex orbital mission. Together with the European Space Agency ESA, NASA sent Cassini-Huygens to the ring planet in 1997. Initially, like most interplanetary missions to the outer solar system, Cassini-Huygens had to be accelerated by close fly-bys at other planets. This was done twice at Venus, once at Earth, and once at Jupiter. This was enough to arrive at Saturn in mid-2004. At the end of the same year, Huygens was dropped off, landed softly on the Moon Titan, and sent images and measurement data from there. Cassini itself moved on elongated ellipses through the Saturn system until the planned crash into the Saturn atmosphere in September 2017. Cassini-Huygens was the only interplanetary enterprise from which no images of a gravity assist at Earth exist. However, the Cassini orbiter has sent us the first images of our home planet from the orbit of an outer, distant planet. From Saturn's distance, the Earth can only move a few degrees away from the Sun. For Cassini's wide-angle camera with a field of view of 3.5°, it was therefore not an easy subject. Solely from For security reasons, it was necessary to avoid letting the instruments look directly into the Sun. However, on September 15, 2006, Saturn itselfcame to the aid of the operators in the control center. In the orbit currently taking place, 2.1 million km away from the probe, the ringed planet obscured our central star and acted as a sunshade. Now the camera could target and photograph the Earth, which was standing directly next to Saturn. From a distance of 1.5 billion km, it appears through Saturn's ring system. The resolution is not sufficient to make out details on it; if one could, one would look at the Atlantic and the west coast of Africa. Upon closer inspection, however, one can see that the Earth apparently has an extension on one side. This is our companion, the Moon. It can actually still be identified from the distance of Saturn. All previously shown images of the Earth were created without the participation of the public. If lengthy planning was associated with it, this was done in expert circles and the results were later presented more or less spectacularly.

Things were different on July 19, 2013. Cassini's task that day was to create a portrait of the entire Saturn system from wide-angle shots. The telecamera was also occasionally used. It has a significantly smaller field of view, but a much higher resolution. The Sun was again shielded by Saturn when the telecamera was pointed at the Earth, 1.45 billion km away. In the wide-angle shot, Earth and Moon form an inseparable bright pair to the right below the ringed planet; the telephoto shot then shows both as clearly separated points of light. At the same time, Messenger also photographed the Earth from its orbit around Mercury. Two portraits of our home were taken that day, taken from different, widely separated positions in the solar system. Anyone who registered NASA's announcement of this event and accordingly stayed under the open sky became part of these recordings as an Earth inhabitant. The day went down in the history of space travel as the "day the Earth smiled" Cassini last had the opportunity to photograph the Earth from a great distance on April 12, 2017, a few months before the end of its mission, which was sealed in September 2017 with Cassini's entry into Saturn's atmosphere. The author Bruno H. Bürgel mentioned in the introduction would probably be surprised at how images of the Earth with its Moon even succeed from the distance of Saturn today.

Sun

For probes that we send towards the center of the solar system to study our central star from as close as possible, the same applies as for missions that we send to our planetary neighbors. They too move in interplanetary space under the gravitational influence of the Sun and follow the laws of gravity. For the first time, the two Helios probes, a joint project of the DFVLR (today DLR) and NASA in the mid-70s, approached the Sun and reached a minimum solar distance of just over 40 million km. Many of the Sun probes launched since then have dedicated themselves to our central star with instrumentation that, like the two Helios probes, did not allow photos of the Earth from interplanetary space. Two current missions, NASA's "Parker Solar Probe" and the "Solar Orbiter" a cooperation between ESA and NASA, are among other things concerned with the particle stream that is constantly emitted by the Sun. This solar wind creates "space weather", when its particles are hitting Earth, they can influence its immediate surroundings.

Both probes are equipped with cameras that allow images of the Earth from interplanetary space. Two years ago, they actually managed to take photos of our home planet that show it together with its neighbors. In particular, the image of the Parker Solar Probe reminds of the family portraits of the solar system that Voyager 1 transmitted to Earth more than 30 and Messenger more than 10 years ago.

225 One of the first images of the Earth from aboard an interplanetary spacecraft. From a distance of 2.6 million km, this mosaic of Earth and Moon was created, placing both celestial bodies side by side in the image. Mariner 10, November 1973. (Photo: NASA, JPL, Northwestern University)

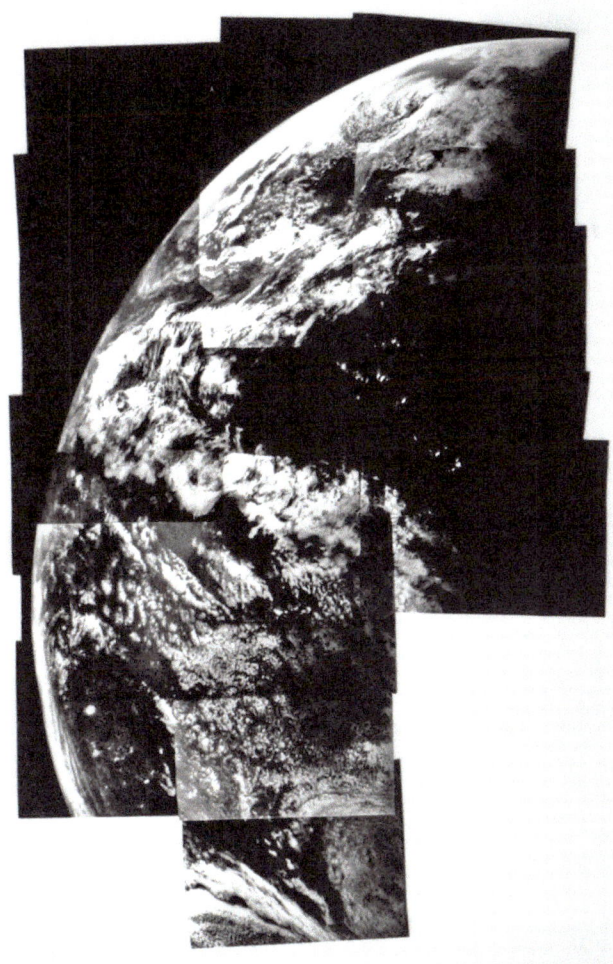

227 Above: Before Messenger approached its planned gravity maneuver of the Earth, a picture of Earth and Moon was taken from a distance of 29.6 million km. Messenger, May 11, 2005. (Photo: NASA, Johns Hopkins University Applied Physics Laboratory, Carnegie Institution of Washington)

226 Left: Shortly after the launch of Mariner 10, its cameras were already switched on to check the effects of an error that had occurred on board. From this, this mosaic of the cloud-covered Earth was created. It proved that the cameras were working as planned and the recordings of Mariner 10's fly-by of Venus and the investigation of Mercury were not endangered. Mariner 10, November 1973. (Photo: NASA)

228 A mosaic of the Earth with its Moon. The image of the Earth comes from the gravity maneuver in August 2005 while the image of the Moon was taken three days earlier. Messenger, July 31, 2005 and August 2, 2005. (Photo: NASA, Johns Hopkins University Applied Physics Laboratory, Carnegie Institution of Washington, Gordan Ugarkovic)

229 Five years later in May 2010, Messenger was 183 million km away from Earth. From this position, an attempt was made to find the Vulcanoids, hypothetical asteroids that are supposed to be located between Mercury and the Sun. In this search, the Earth-Moon system appeared as closely spaced bright points of light in the constellation Libra. The slightly brighter star to the right of Earth and Moon is δ Scorpii, North points downwards. Messenger, May 6, 2010.

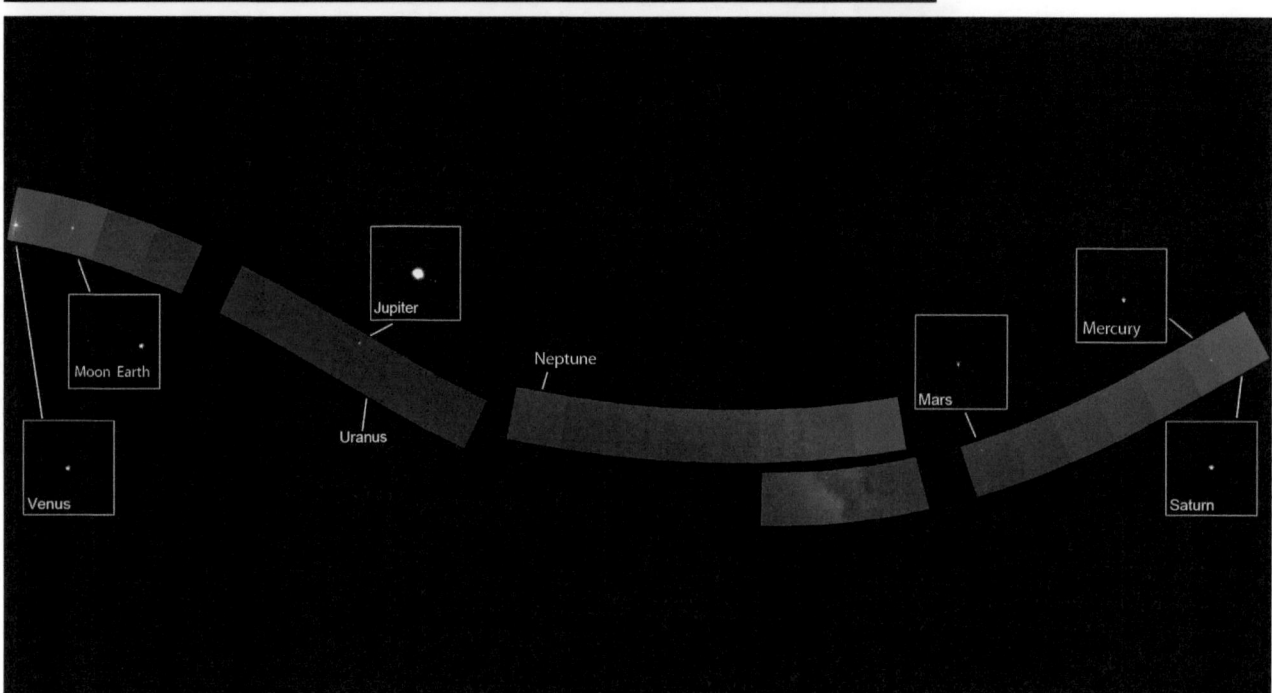

230 In November 2010, Messenger attempted a "family portrait" of the solar system, similar to that of Voyager 1 in 1990. Only Uranus and Neptune are missing due to their great distance of 3 and 4.4 billion km respectively. The Earth shows its Moon, as does Jupiter some of its brightest Galilean Moons. Messenger, November 2010. (Photo: NASA, Johns Hopkins University Applied Physics Laboratory, Carnegie Institution of Washington)

8. Oktober 2014 09:18:49   8. Oktober 2014 10:08:49

231 On October 8, 2014, a total lunar eclipse occurred, which could be followed on Earth mainly from East Siberia across the Pacific region to the west of North America. Messenger was already orbiting Mercury, but from this distance it was also able to document the disappearance of the Moon. In the left picture, you can see the Moon illuminated by the Sun standing next to the Earth. In the right picture 50 minutes later, the Earth's shadow covers the Moon and makes it disappear for Messenger. Messenger, October 8, 2014. (Photo: NASA, Johns Hopkins University Applied Physics Laboratory, Carnegie Institution of Washington)

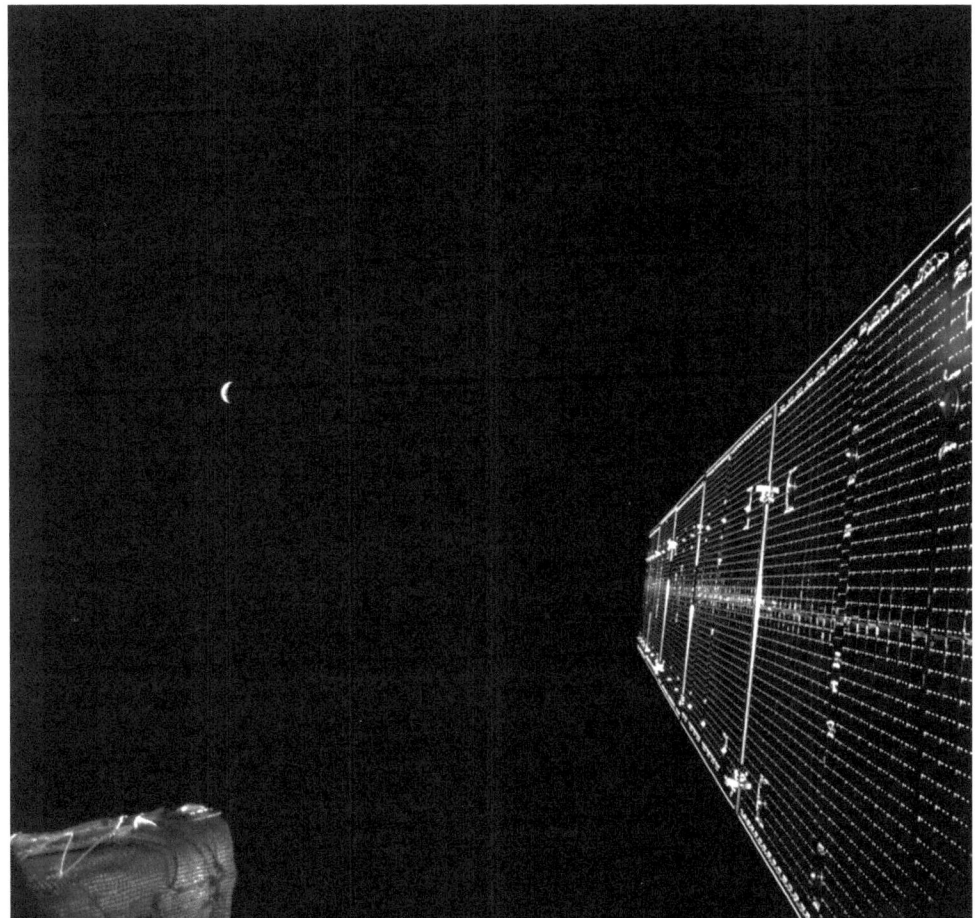

232 Flights to Mercury are rare. After Messenger with a launch in 2004, it took another 14 years until Europe, in cooperation with Japan, launched the probe "BepiColombo", named after Guiseppe Colombo, an Italian spacecraft expert. It will not enter an orbit around Mercury until 2025. Until then, numerous Gravity Assist maneuvers are planned. The first led BepiColombo past the Earth in April 2020. As the probe moved away again, it bid farewell to our significantly smaller home planet from a distance of 540,000 km. BepiColombo, April 10, 2020. (Photo: ESA/ BepiColombo/MTM)

233 Right: For the flight to Mars, no trajectory changes at other celestial bodies in the solar system are actually required. It was different with the Japanese Mars probe Nozomi. Since its carrier rocket did not provide the necessary thrust, it was supposed to make two Gravity assist maneuvers at the Moon to put it on the right path. Unfortunately, the second one failed, which ultimately led to the failure of this mission. Nevertheless, Nozomi stands for a special feature: Although cameras were part of the equipment of all Mars probes, it took more than three decades since the successful fly-by of Mariner 4 in July 1965 for one of them to deliver a photo of the Earth. This was Nozomi with a portrait of Earth and Moon shortly after launch. Our home planet was 170,000 km away at the time of the recording, the Moon 540,000 km away. Nozomi, July 18, 1998. (Photo: JAXA, ISAS, Nozomi MIC Team)

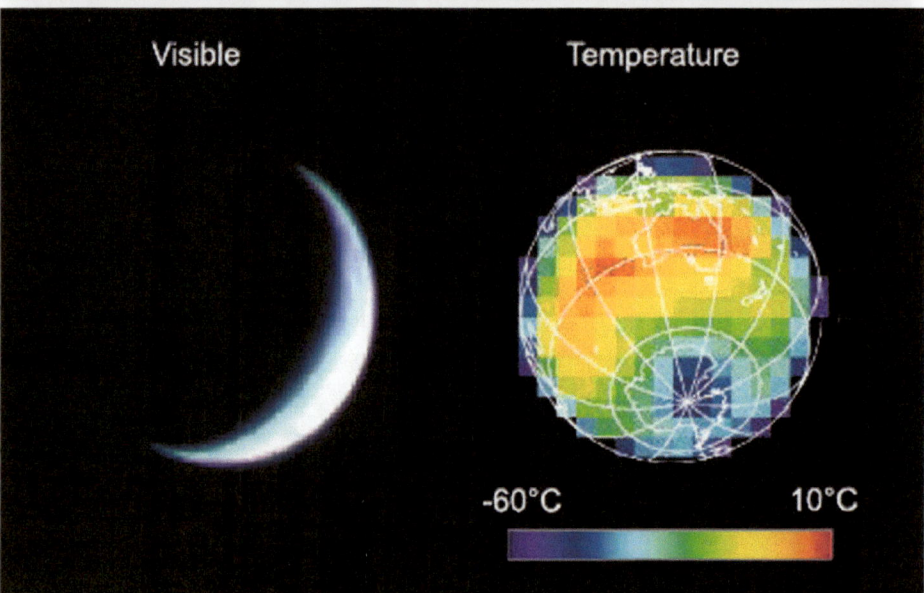

234 After the NASA probe Mars Odyssey had covered about 3.5 million km on its way to the Red Planet, it pointed its camera system at the Earth. The left image shows it as a narrow crescent in bright visible light. The right image illustrates the temperature distribution on Earth during this recording, derived from the data of the infrared sensor. The coldest spot is found in blue at the bottom right. This is Antarctica, while red tones above it indicate the warmest region with Australia. Mars Odyssey, April 19, 2001. (Photo: NASA, JPL, Arizona State University)

235 The image of the Earth with the Moon (right) from Mars Odyssey dates from the same day. Here, both celestial bodies are exceptionally not shown in visible light, but also in infrared wavelengths, which allow conclusions about the prevailing temperatures. Again, the dark cold South Pole with the least thermal emission can be seen, while Australia can be identified as a hot, bright spot above it. Mars Odyssey, April 19, 2001. (Photo: NASA, JPL, Arizona State University)

236 In May 2003, the Mars Global Surveyor, which had been orbiting Mars since 1997, was able to capture a conjunction of Earth and Jupiter, i.e., an apparent encounter of both planets. The distance between Mars and Earth was 139 million km, that between Mars and Jupiter 944 million km. The camera only recorded grayscale images. For the image shown here, Earth and Jupiter were colored as realistically as possible based on earlier data of the probes Mariner 10 (Earth) and Cassini (Jupiter). In the two enlarged sections, the Moon can be seen next to the Earth (top) and Jupiter is framed by three of its four Galilean moons (bottom) with Callisto, Ganymede, and Europa (from left to right). Io was behind Jupiter's disk at the time of the recording. The brightly illuminated half-Earth was facing the camera with both American continents. Mars Global Surveyor, May 8, 2003. (Photo: NASA, JPL, Malin Space Science Systems)

237 Shortly before Earth and Mars approached each other in 2003 to only 55.8 million km, the smallest mutual distance in a long time, Europe sent its first Mars probe Mars Express on its way. The landing device Beagle 2 failed to land on the Martian surface as planned, but Mars Express has been able to successfully carry out its tasks since the end of 2003. The high-resolution stereo camera was tested for the first time when Mars Express had already covered 8 million km one month after launch and took a picture of the Earth with its Moon from this distance. Mars Express, July 3, 2003. (Photo: ESA/DLR/Free University of Berlin)

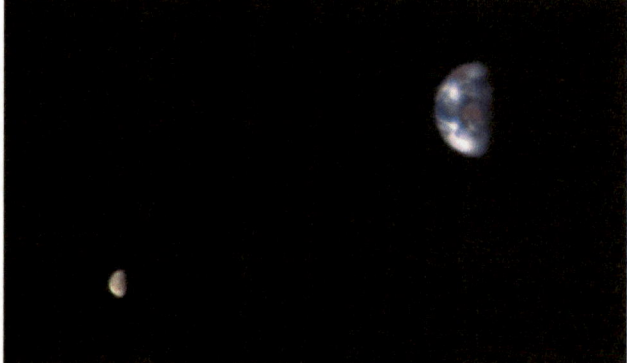

239 Nine years after the depiction of Earth and Moon shown in the previous image, both celestial bodies reappeared in the field of view of the Mars Reconnaissance Orbiter, but now significantly further away at 205 million km. The reddish-brown spot roughly in the center of the half-Earth is Australia. Mars Reconnaissance Orbiter, November 20, 2016. (Photo: NASA, JPL-Caltech, University of Arizona)

238 The camera system on the Mars Reconnaissance Orbiter currently provides the highest resolution images from Mars orbit. In October 2007, the Earth-Moon system came into its field of view. This image from a distance of 142 million km is significantly more detailed than the photo taken from 139 million km in figure 236. The technology of the equipment of interplanetary probes has obviously made continuous progress. Mars Reconnaissance Orbiter, October 3, 2007. (Photo: NASA, JPL-Caltech, University of Arizona)

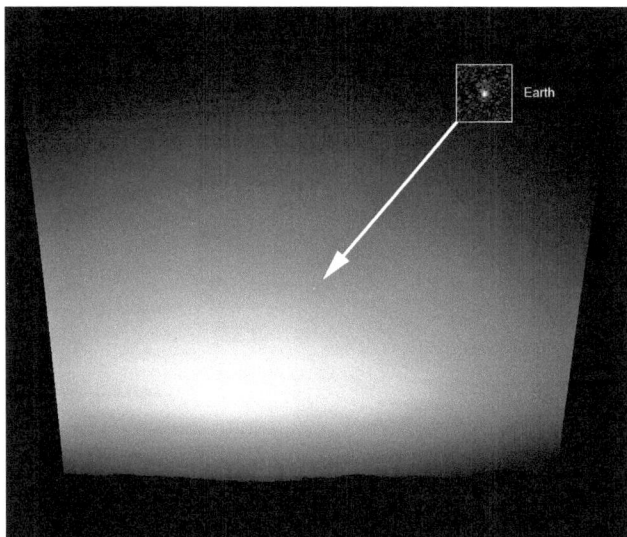

240 Depictions of Earth from the surface of another celestial body have a special charm. Previous examples only concern the Moon, either taken by landed stationary and mobile probes or seen by astronauts during the short phase of manned Moon landings. For the first time, an image of Earth was taken from a celestial body beyond the Moon by the mobile Mars Exploration Rover named "Spirit" In March 2004, Earth appeared as a bright point above the Martian horizon. This image is a mosaic of images from Spirit's navigation camera and its panoramic camera. Mars Exploration Rover Spirit, March 11, 2004. (Photo: NASA, JPL, Cornell, Texas A&M)

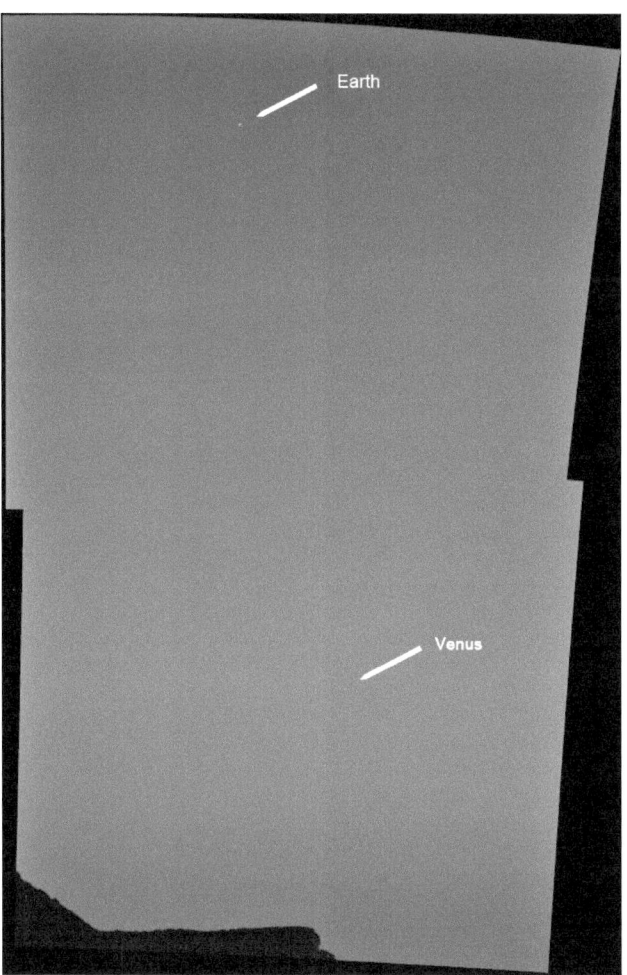

242 The Mars Science Laboratory, named "Curiosity", managed to take a photo of Earth with Venus, its closest planetary neighbor, in June 2020. Both planets, Earth above and Venus further below, appear as point-like light sources in the Martian sky above the rock outcrop named Tower Butte. Curiosity, June 5, 2020. (Photo: NASA, JPL)

241 Left: Earth and Moon were visible from the surface of Mars in January 2014. "Curiosity" has been moving across Mars since the end of 2011 when its mast camera captured both celestial bodies above the silhouette of the Martian horizon. Curiosity, January 31, 2014. (Photo: NASA, JPL)

243 NEAR Shoemaker, sent to the asteroid Eros in 1996, needed a gravity assist maneuver at Earth to reach its destination. This took place in January 1998. After the fly-by, NEAR Shoemaker had already moved 400,000 km away, Earth appeared together with its Moon. NEAR Shoemaker, January 23, 1998. (Photo: NASA, NEAR Spacecraft Team, Johns Hopkins University Applied Physics Laboratory)

244 Hayabusa also, sent into interplanetary space with the ambitious mission to bring samples of the asteroid Itokawa back to Earth, took the route via our home planet. In May 2004, the probe passed Earth, documented here with a picture of the Atlantic. To the west are North and South America, to the east Africa. Hayabusa, May 18, 2004. (Photo: JAXA, ISAS)

245 Hayabusa-2 also passed Earth, like its predecessor It closely examined the asteroid Ryugu and returned to Earth with samples from it, an Earth-crossing asteroid. The sight of our home planet from a distance of 340,000 km during the flyby in 2015 resembled that of NEAR Shoemaker in 1998 with the prominently visible Antarctica. Hayabusa-2, December 4, 2015. (Photo: JAXA)

246 Shortly before, as Hayabusa-2 approached Earth for the gravity assist, its camera looked at the Earth-Moon system from a distance of 3 million km. Hayabusa-2, November 26, 2015. (Photo: JAXA)

247 The European comet probe Rosetta performed a true interplanetary dance during its 10-year flight to the comet Churyumov-Gerasimenko. Three gravity maneuvers at Earth and an additional swing-by at Mars were necessary to swing Rosetta into an orbit around the comet. The second swing-by maneuver at Earth took place in November 2007. During this second flyby of Rosetta created this view of Southeast Asia, Australia, and the eastern Pacific. Rosetta, November 15, 2007. (Photo: ESA ©2007 MPS for OSIRIS Team MPS/ UPD/LAM/IAA/ RSSD/INTA/UPM/DASP/ IDA, Gordan Ugarkovic)

248 Rosetta's second swing-by maneuver at Earth in November 2007 took the comet probe over its night side, capturing Southeast Europe, North Africa, and southern Asia with its nighttime lighting from a distance of 80,000 km. Somewhat later, 75,000 km away, a view of the southern hemisphere was taken, with the part of the southern hemisphere illuminated by the Sun one month before the winter solstice. Both images are processed into a mosaic here. Rosetta, November 13, 2007. (Photo: ESA ©2007 MPS for OSIRIS Team/MPS/UPD/ LAM/ IAA/RSSD/INTA/UPM/ DASP/IDA)

249 For the last time, Rosetta passed by Earth in November 2009. The shortest distance during this flyby was 2481 km. As it approached our home planet, the photo of the narrow Earth crescent (left) was taken. As it moved away, Rosetta looked at a significantly fuller Earth (right). Rosetta, November 12, 2009. (Photo: ESA ©2009 MPS for OSIRIS Team MPS/UPD/LAM/IAA/RSSD/INTA/UPM/DASP/IDA, Gordan Ugarkovic)

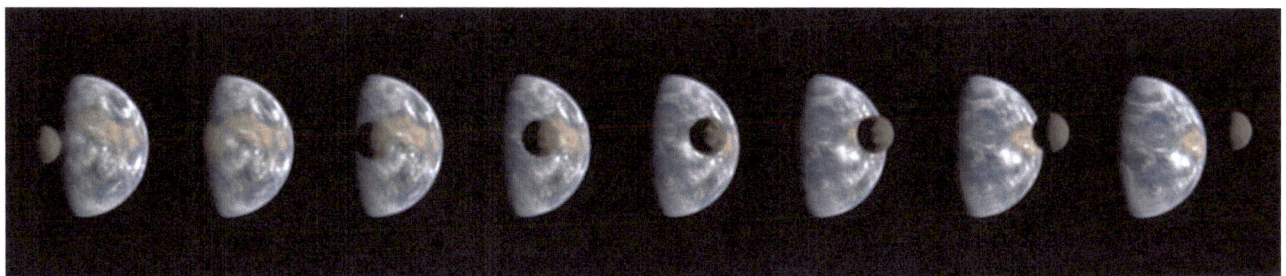

250 From very different distances, the Deep Impact probe, now traveling under the name EPOXI after a successful mission at Comet Temple 1, observed our Earth. 50 million km away from it, it pointed its camera at it and the passing Moon with the aim of understanding what Earth-like planets look like from a great distance. Deep Impact/EPOXI, May 29, 2008. (Photo: NASA, GSFC, JPL-Caltech, UMD)

251 The OSIRIS-REx probe is on its way to the asteroid Bennu. In September 2017, it gained momentum from Earth. Three days after the closest approach to Earth, the probe was already 1.3 million km away. Now our companion, the Moon, also appeared in the field of view of its navigation camera. OSIRIS-REx, September 25, 2017. (Photo: NASA, GSFC, University of Arizona)

252 Another week later, OSIRIS-Rex had now put more than 5 million km between itself and the Earth, creating another image of the Earth-Moon system. OSIRIS-REx, October 2, 2017. (Photo: NASA, OSIRIS REx Team, University of Arizona)

253 By the beginning of 2018, already 64 million km had been covered when the navigation camera took this black and white image of Earth and the Moon standing to its right. Earth and Moon are in the constellation Cetus. The bright star diagonally below the Earth is Menkar (α Ceti) and to the right of it is the well-known variable star Mira (o Ceti), which reached its greatest brightness shortly after the time of the recording. In the top left, you can see the open star cluster of the Pleiades, the Seven Sisters, in the constellation Taurus. OSIRIS-REx, January 17, 2018. (Photo: NASA, GSFC, University of Arizona, Lockheed Martin)

255 The last available photo of Earth from a space probe dedicated to the exploration of asteroids. It was taken from a distance of 620,000 km, captured by the Lucy probe on its way to the Trojan asteroids. Lucy, October 22, 2022. (Photo: NASA, GSFC, SwRI, Johns Hopkins University Applied Physics Laboratory, Tod R. Lauer/NOIRLab)

254 In December 2018, OSIRIS-REx reached the asteroid Bennu. From a distance of only 43 km, Bennu is overexposed in the right half of the image. Earth and Moon on the lower left, on the other hand, appear significantly dimmer – no wonder at a distance of 114 million km. OSIRIS-REx, December 19, 2018. (Photo: NASA, GSFC, University of Arizona, Lockheed Martin)

256 In the 70s, NASA sent four probes to the outer solar system and beyond. Contact is still maintained with Voyager 1 launched in 1977. Considering the distance covered by both spacecraft so far, a portrait of Earth and Moon was created for the first time in September 1977, with its brightness slightly increased, practically on the terrestrial doorstep, namely from a distance of only 11.7 million km. The quality of the image taken by Voyager 1 already hinted at the great results to be expected from the planned fly-bys of the gas and ice planets. Voyager 1, September 18, 1977. (Photo: NASA, JPL)

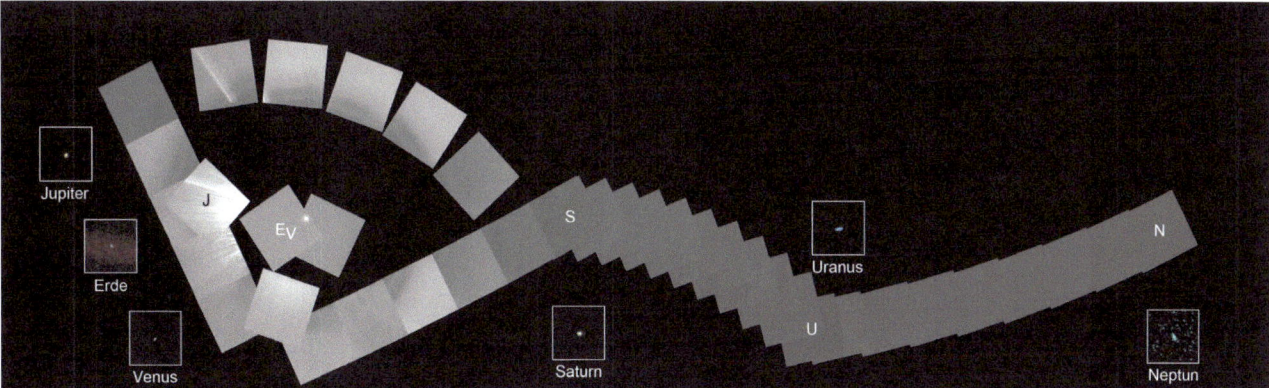

257 More than 12 years later, Voyager 1 was 32° above the ecliptic plane and about six billion km away, already at a medium distance from the dwarf planet Pluto. A look back allowed a "family portrait" of the entire solar system. It consists of several individual shots that were assembled into a mosaic. At the center is the Sun, whose brightness in some of the individual shots produces strong scattered light effects. Mercury and Mars are missing, as Mercury was unfavorably close to our central star and Mars was too heavily overlaid by solar scattered light. Earth ("E") and Venus ("V") are also found directly next to the Sun. In the individual shot of our Earth (see also next page), it appears as a "pale blue dot", within a strip of scattered Sunlight. Voyager 1, February 14, 1990. (Photo: NASA, JPL)

258 In the more than 30-year-old original individual shot (right) of Voyager's family portrait of the solar system, shown on the previous page, the Earth stands as a "pale blue dot" in a bright solar scattered light strip. Next to it, further colored effects of this kind are shown. Although no details can be seen of our home planet in this shot, this photo has achieved cult status - there is no image of the Earth from a similar distance of 6.4 billion kilometers. Moreover, it is one of the last photos that Voyager 1 sent to Earth. 34 minutes after this picture, the cameras of Voyager 1 were permanently switched off. For the 30th anniversary of this photo, NASA has reprocessed it using modern image processing techniques (below). The color shifts are corrected and the noise has been minimized, so the scene now appears in natural colors. Voyager 1, February 14, 1990. (Photo: NASA, JPLCaltech)

259 After the close Earth fly-by in early December 1990, the Jupiter probe Galileo had already moved more than two million km away from Earth when the probe photographed the visible hemispheres at intervals of six hours. South America (top left), India and Australia (top right), the Pacific (bottom left) and Africa (bottom right). Galileo, December 11, 1990. (Photo: NASA, JPL)

260 Two years later, Galileo passed by Earth again. Now already 6.2 million km away, the farewell photo of Earth with its Moon was taken. Galileo, December 16, 1992. (Photo: NASA, JPL, University of Arizona, Gordan Ugarkovic)

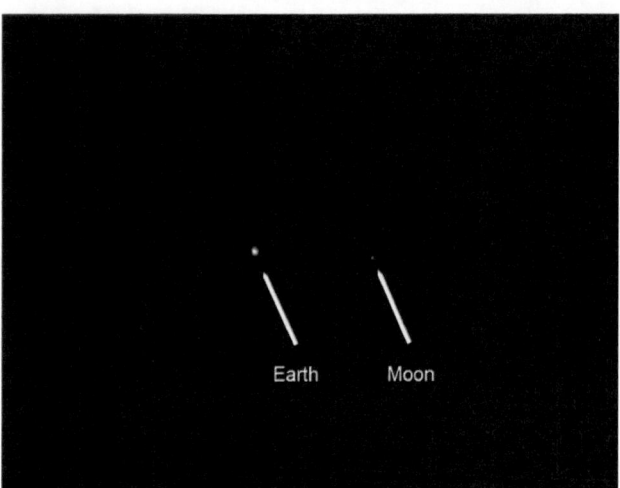

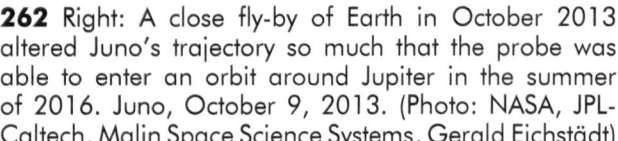

261 A rather unspectacular shot of Earth and Moon, transmitted by Juno. Three weeks after its launch in 2011, this second Jupiter probe from NASA had already put 9.7 million km between itself and Earth. From this position, it photographed the Earth-Moon system. What's interesting about this photo is rather the timing. It happened almost exactly 34 years after Voyager 1 had first depicted the Earth from a similar interplanetary distance. Juno, August 26, 2011. (Photo: NASA, JPL-Caltech, Malin Space Science Systems)

262 Right: A close fly-by of Earth in October 2013 altered Juno's trajectory so much that the probe was able to enter an orbit around Jupiter in the summer of 2016. Juno, October 9, 2013. (Photo: NASA, JPL-Caltech, Malin Space Science Systems, Gerald Eichstädt)

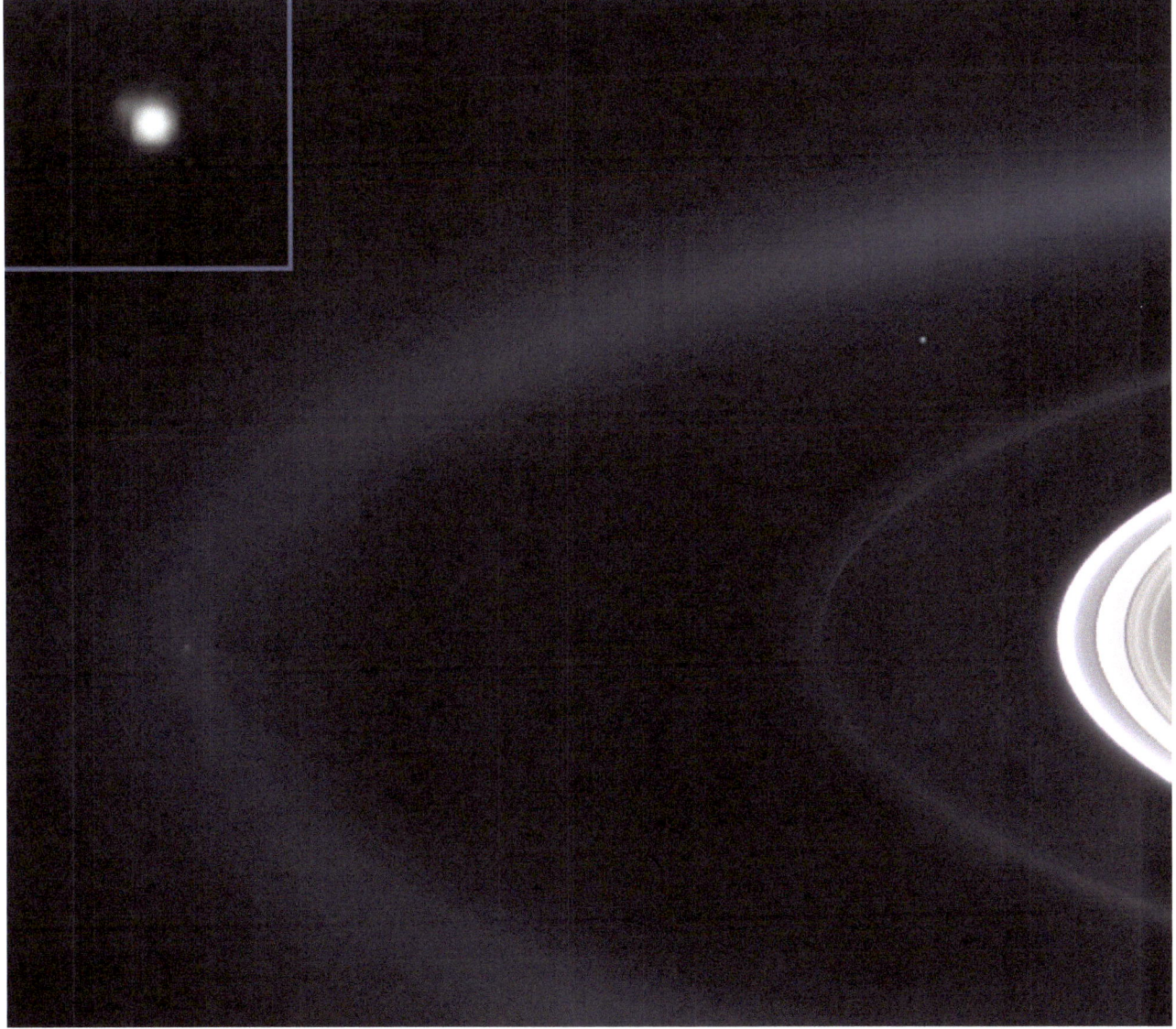

263 Only Voyager 1 was further away from Earth when taking a picture than the Saturn probe Cassini in September 2006. From a distance of 1.5 billion km, the Earth is seen here as the brightest point source between the outer bluish diffuse E-ring and the narrow G-ring in the upper right half of the picture. On the right edge of the picture, Saturn's bright main rings can be seen. In the enlarged monochrome image in the upper left, the Moon can be made out as an elongated appendage. If one could have seen the surface of the Earth from Saturn's distance, the Atlantic Ocean and the west coast of Africa would have been in the field of view. This shot was only possible because from Cassini's position Saturn obscured the Sun. Cassini, September 15, 2006. (Photo: NASA, JPL, Space Science Institute)

264 Three images of Saturn with its ring system. In all three, the Earth can be seen. From wide-angle shots at a time in July 2013 from a distance of 1.44 billion km, when Saturn obscured the Sun, the entire Saturn system was depicted (above). The Earth appears to be, together with the unresolved Moon, through the inner edge of the diffuse E-ring. In the picture on the bottom left, the Earth is seen as the brightest point source right below the narrow G-ring. The further inside F-ring is overexposed, just like the narrow illuminated edge of Saturn, which is interrupted by the shadows of the rings. Finally, the Moon also appears in the image of the telecamera next to the brighter Earth (bottom right). Now the Earth is also clearly recognizable as a "pale blue dot". Cassini, July 19, 2013. (Photo: NASA, JPL, SpaceScience Institute)

265 In April 2017, there was again the opportunity, during an eclipse of the Sun by Saturn, to photograph the Earth from a great distance of 1.4 billion km. In the image above, it shines as a bright point below the A-ring. In the ring system, the Keeler and Encke divisions are clearly visible. The bright ring below the Earth is the F-ring. If you enlarge the area around the Earth in this photo, the Moon also becomes more clearly visible (below). Cassini, April 12, 2017. (Photo: NASA, JPL-Caltech, Space Science Institute)

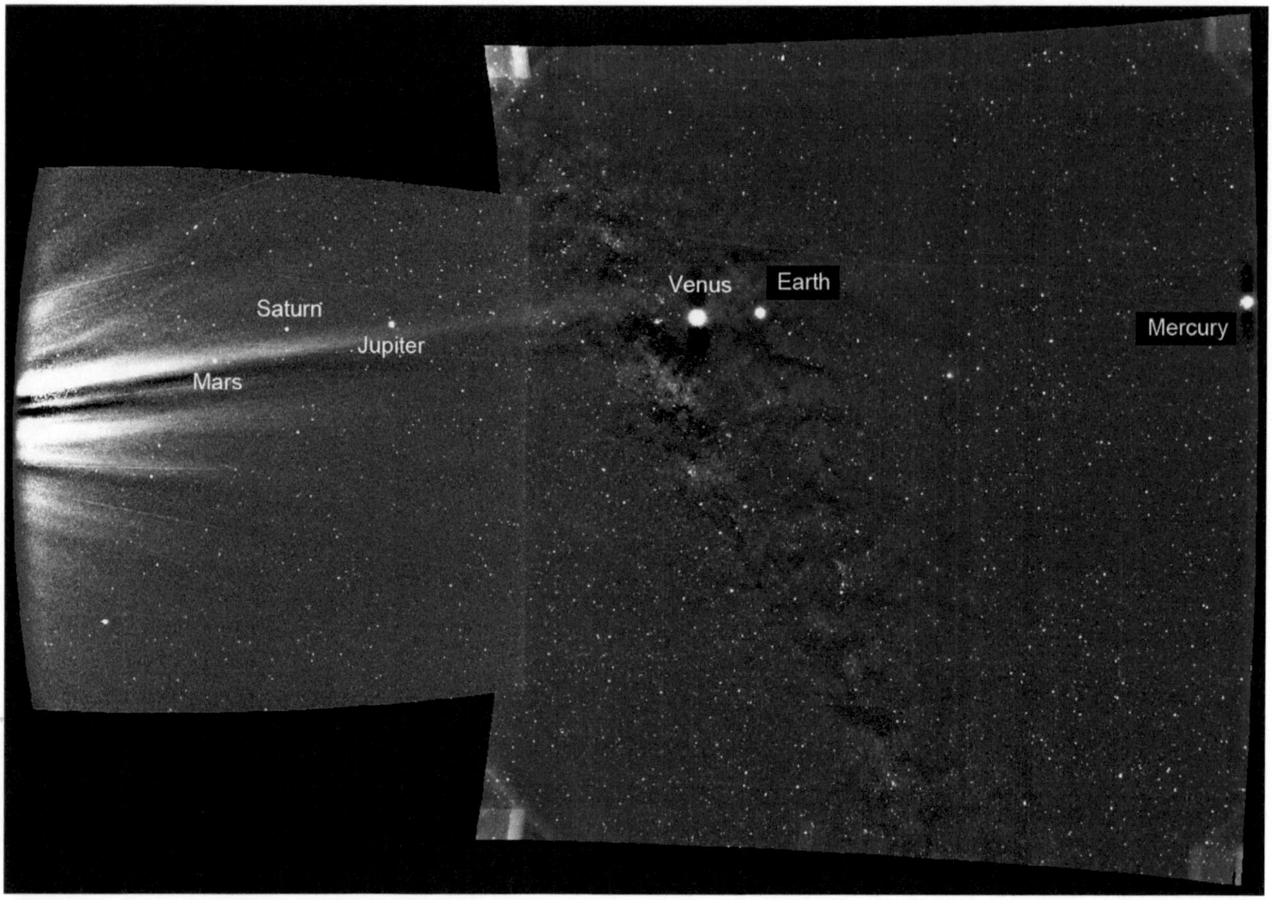

266 When the Parker Solar Probe was only 19 million km away from the Sun in June 2020, its wide-angle camera saw all planets except Uranus and Neptune lined up to the right of the Sun. Venus and Earth are just pointing towards the center of our Milky Way. Parker Solar Probe, June 7, 2020. (Photo: NASA, Johns Hopkins University Applied Physics Laboratory, Naval Research Laboratory, Guillermo Stenborg, Brendan Gallagher)

267 Months later, the Solar Orbiter also spotted three of these planets. The Earth was at a distance of 251 million km while Venus at 48 million km was much closer and Mars was already 332 million km away. Solar Orbiter, November 18, 2020. (Photo: Solar Orbiter/SoloHI Team/ESA & NASA, U.S. Naval Research Laboratory)

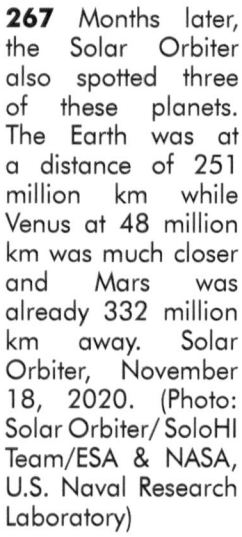

Epilogue

We have now reached the end of the journey. Initially, it took us great efforts to even see the curvature of the Earth's horizon. Later, Earth appeared to us as a wonderful "blue marble" until it finally no longer revealed the unique details of its surface in the outer solar system and only presented itself as a " pale blue dot".

When the record set by Voyager 1, transmitting a picture of the Earth from more than six billion km away, can be surpassed remains to be seen. Currently, the NASA probe New Horizons has passed the dwarf planet Pluto and the trans-Neptunian object Arrokoth in the Kuiper belt. This belt houses small icy objects that can reach Pluto dimensions in size and originate from the early days of the solar system. Its extent reaches up to a solar distance of about 7.5 billion km. This corresponds to the current distance of New Horizons. After passing the Kuiper belt, the probe will leave our solar system like the two Pioneer and Voyager missions before it. However, New Horizons has not transmitted a " blue dot" like Voyager 1. Whether it will ever happen again is literally written in the stars.

What was denied to earlier generations, namely to see the Earth as a planet, as part of the solar system and thus as part of the universe, we have been able to witness in recent decades through the development of space travel. Only fiction and artistic skill allowed the Earth to appear in the sky in the pre-space travel era, seen from spaceships or from the surface of alien worlds. Some of these considerations were too fantastic, some came quite close to reality. Bürgel's statement mentioned in the introduction, that one can no longer recognize the Earth from beyond Saturn because it is hidden in the light of the in the light of the Sun, is basically correct. He probably just couldn't imagine how we could position a probe in orbit around Saturn so that the ringed planet serves as Sun protection, or how we could use image processing methods to locate the faintly glowing Earth even in solar scattered light. In the early days of space travel, there was even the idea that pictures of the Earth from space, especially views of a fully illuminated Earth, would influence our view of things with positive effects on world events. Unfortunately, this idea did not come true. Some pictures of the Earth became symbols in the early days of the environmental movement. For the first "Earth Day" on April 22, 1970, a day to focus on environmental issues, the later "blue marble" photo from Apollo 17 became an icon. The pictures from space impressively showed how beautiful, but also vulnerable our planet is, the only home in which we can exist.

Only the scientific and technical progress of the last decades made it possible to see the Earth as part of the universe. Developments in related fields led to a new class of objects coming into focus in astronomy during the same period and becoming the subject of intensive research - the exosolar planets. It is an interesting temporal coincidence how we try to represent the Earth as a planet in space travel on the one hand, and to see how it presents itself from the greatest possible distances, and on the other hand in astronomy we overcome the distance to foreign solar systems with the help of ever more powerful telescopes in order to be able to depict planets orbiting a central star as points of light. Only when we "see"; these planets on our detectors will we be able to examine them more closely and pursue one of the big questions of today's astronomy, whether one of them is a "second Earth". Since the discovery of the first exoplanet in 1995, numerous others have been found and classified. A selection effect - the detection methods often favor very massive or close to the central star standing planets - leads to the fact that we only slowly advance into the area of a possible "second Earth". When the first exoplanet was detected, the picture from Voyager-1 was already available. Similar pictures, but not only bridging 6 billion km, but light years, are the goal in astronomical research. When we examine the

M. Gottwald, *The Earth*, https://doi.org/10.1007/978-3-662-69633-0

light of the then visible exoplanets more closely, we will know whether it is also a "pale blue dot", that is a world that also exhibits habitability. Until we have found a "second Earth" we only know of one place in the universe that allows life as we know it, namely our home planet. And it is very likely that we are bound to this for the duration of our existence.

Used Abbreviations

AGU American Geophysical Union
CLEP China's Lunar Exploration Program
CNSA China National Space Administration
DLR German Aerospace Center
ESA European Space Agency
EUMETSAT European Organisation for the Exploitation of Meteorological Satellites
GSFC Goddard Space Flight Center
ISRO Indian Space Research Organisation
JAXA Japan Aerospace Exploration Agency
JMA Japanese Meteorological Agency
JPL Jet Propulsion Laboratory
LPI Lunar and Planetary Institute
NASA National Aeronautics and Space Administration
NOAA National Oceanic and Atmospheric Administration
NRO National Reconnaissance Office
USGS United States Geological Survey
UTC Coordinated Universal Time (MEZ = UTC+1 hour)

Index